Copyright © 2023 by Mark Newman

All rights reserved. No part of this book may be reproduced or used in any manner without written permission of the copyright owner except for the use of quotations in a book review.

For more information:

mark@howthefouriertransformworks.com

First edition January 2023

ISBN 979-8-3739-2403-0 (paperback)

howthefouriertransformworks.com

Table of Contents

Preface 1

Chapter 1: From Fourier Series to Fourier Transform 2

    Infinite time 2

    Continuous frequency... 6

    Interference 7

    Beat frequencies 8

    Extending the silence.. 10

    Bandwidth... 13

    Coping with infinity 14

Chapter 2: The Discrete-Time Fourier Transform 16

    A brief history of sound-recording media 16

    The problem with the Fourier Transform 17

    Comparing the Fourier Transform and the DTFT 18

        The complex exponential.. 18

        A discrete-time signal instead of a continuous one................................... 18

        Summation instead of integration 19

    Calculating the DTFT – a numerical example 20

    The frequency periodicity of discrete-time signals 22

    Aliasing and the Nyquist Rate 24

    Infinities remaining in the DTFT 26

Chapter 3: The Discrete Fourier Transform 27

    Comparing the DTFT and DFT 27

Testing a finite number of frequencies in the DFT 28

The complex exponential 28

Looking at the signal for a finite amount of time 29

Frequency resolution in the DFT 30

    Calculating the frequency step size 31

Calculating the DFT – a numerical example 31

    Calculating the DFT of frequency index $k=0$ 33

    Calculating the DFT of Frequency index $k=1$ 33

    Calculating the DFT of frequency index $k=2$ 34

    Calculating the DFT of frequency index $k=3$ 34

    Calculating the DFT of frequency index $k=4$ 35

    Calculating the DFT of frequency index $k=5$ 35

    Calculating the DFT of frequency index $k=6$ 36

    Calculating the DFT of frequency index $k=7$ 36

    Calculating the DFT of frequency index $k=8$ 37

    Calculating the DFT of frequency index $k=9$ 37

    The DFT for our signal 38

Long signals and short DFTs 39

Chapter 4: Windowing Functions 40

    Why do we need windowing functions? 41

        What is spectral leakage? 42

        Changing the block size 43

    Different windowing functions 45

        Hann windowing function 45

Hamming windowing function 46

Bartlett windowing function 48

Tukey windowing function 49

Choosing a windowing function 50

Chapter 5: The Fast Fourier Transform 51

Repeating calculations 51

Divide-and-conquer 56

Sorting an array of numbers 56

Divide-and-conquer in the FFT 60

Divide stage 60

Conquer stage 61

The butterfly diagram 66

The 2-point butterfly 66

The 4-point butterfly 69

Twiddle factors 75

The problem with the FFT 75

What are twiddle factors? 79

Calculating twiddle factors 80

Calculating the second group of 4-point DFTs 83

Calculating the FFT – a numerical example 88

Calculating the DFT of frequency index k=0 88

Calculating the DFT of frequency index k=4 91

Calculating the DFT of frequency index k=1 93

Calculating the DFT of frequency index k=5 96

- Calculating the DFT of frequency index $k=2$ 97
- Calculating the DFT of frequency index $k=6$ 99
- Calculating the DFT of frequency index $k=3$ 101
- Calculating the DFT of frequency index $k=7$ 103

Making sense of the results 104

- Frequency 105
- Magnitude 105
- Phase 105

The FFT for our signal 107

Chapter 6: Fourier's Legacy 109

Preface

Fourier's discovery that it is possible to build complex signals out of basic sine waves revolutionized the way we look at data. In today's digital age, understanding the data we collect is vital. Fourier's idea was years ahead of its time, and is in use now more than ever before.

In my previous book, *How the Fourier Series Works*, we studied Fourier's original theory, published in 1822. The Fourier Series itself can only model repeating signals. It took the further work of the German mathematician Peter Gustav Lejeune Dirichlet to expand the capabilities of the Fourier Series so that it could model non-repeating signals.

In this book, we'll follow the development of the Fourier Transform from Fourier's original idea, through Dirichlet's changes, and into the digital age, when we use sampled signals, all the way to the Fast Fourier Transform, which is used so widely today.

Although the Fourier Transform has become the Swiss army knife of digital signal processing, lying at the heart of most of the electronic devices we use nowadays, it is not all-powerful. Yes, it can model more of the signals one is likely to meet in the real world, but it is not without its limits.

By looking not only at the changes Dirichlet made but also at why they were necessary, this book aims to afford you a deeper understanding of the Fourier Transform: how it works, how to use it, and what it can and can't do.

There is a standard way of teaching mathematical concepts in schools and universities that I have always found challenging. When I didn't understand something, my teachers would try to explain it to me in terms of mathematical formulae.

Maybe, like me, you need a more intuitive way of understanding. Therefore, my purpose in writing this book is to explain the Fourier Transform to you in a more comprehensive, intuitive, and visual way.

Chapter 1: From Fourier Series to Fourier Transform

In 1822, Jean-Baptiste Joseph Fourier published his groundbreaking memoir, Théorie Analytique de la Chaleur (The Analytical Theory of Heat). In it, he proposed that:

> **Any function of a variable, whether continuous or discontinuous, can be expanded as a series of sines of multiples of the variable.**

Although this memoir is highly regarded now, when Fourier originally presented it, eminent mathematicians like Joseph-Louis Lagrange and Pierre-Simon Laplace challenged his ideas. What was it they were objecting to?

Fourier claimed that *any* function could be expanded as a series of sines. As we saw in my previous book, a function is another name for a signal. So, according to Fourier, it should be possible to build *any* signal out of a series of sinusoids, or sine functions. However, this is not the case. The Fourier Series can model only repeating signals. It took the further work of the German mathematician, Peter Gustav Lejeune Dirichlet, to expand the capabilities of the Fourier Series so that it could model non-repeating signals.

Let's look at the changes which Dirichlet made and why he made them, turning the Fourier Series into the Fourier Transform. The variable *f* in the formula represents the frequency.

$$c_n = \int_{-\frac{P}{2}}^{+\frac{P}{2}} x(t) \cdot e^{-i2\pi f_n t} \cdot dt \qquad X(f) = \int_{-\infty}^{+\infty} x(t) \cdot e^{-i2\pi f t} \cdot dt$$

Equation 1 – The Fourier Series Equation   Equation 2 – The Fourier Transform Equation

Both Equation 1 and Equation 2 look superficially similar. However, Dirichlet made two important changes.

Infinite time

The first thing Dirichlet changed was the limits of the integral. The limits of the Fourier Series integral are between $-P/2$ and $+P/2$, while the limits of the Fourier Transform integral are between $-\infty$ and $+\infty$. What does this mean?

Integration means finding the area under the graph produced by the function within the integral. The limits of the integral tell us how much of the graph to look at.

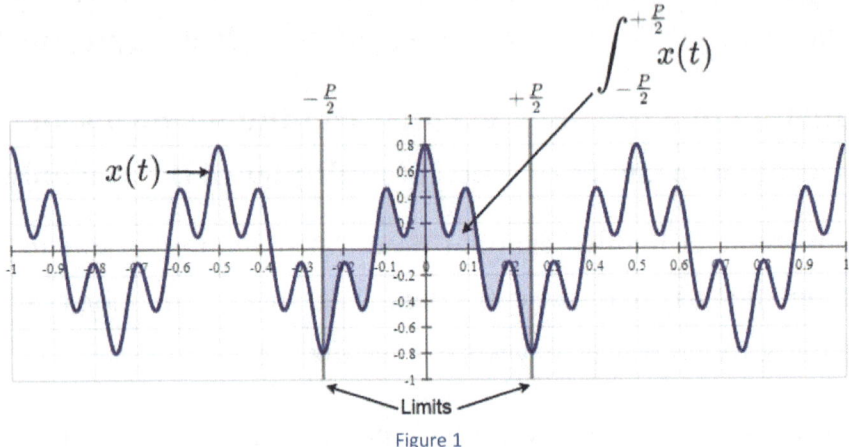

Figure 1

In the Fourier Series, we are looking at the signal for only a limited amount of time, represented by the blue-colored area in Figure 1. The variable P stands for the period of one cycle of the signal. Only repeating signals have a cycle. Therefore, only repeating signals can be represented as a series of sines.

However, most of the useful signals we encounter do not repeat themselves. So Fourier's claim that *any* signal can be represented as a series of sines cannot be true.

In contrast, Equation 2, the Fourier Transform equation, looks at the entire signal. The reason for this is that Dirichlet wanted to model signals that never repeat themselves and therefore have no cycle. "Never" means the whole of time, so we need to look at the signal from the beginning of time to the end of time. And since time is infinite, we need to look at the signal from $t=-\infty$ to $t=+\infty$.

This immediately sets a limit on the type of signal that the Fourier Transform can model. For example, it cannot model a signal like the one shown in Equation 3.

$$x(t) = e^t$$

Equation 3

Why is this?

To answer this question, we first need to see what Equation 3 looks like on a graph, like the one in Figure 2.

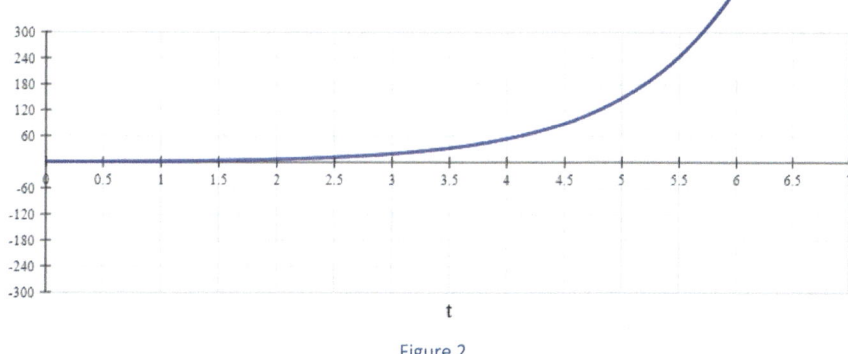

Figure 2

To find the sine waves present in a signal, Fourier uses the following algorithm:

1. Multiply the signal by a cosine wave at the frequency we are looking for.
2. Measure the area under the multiplied cosine signal graph.
3. Multiply the signal by a sine wave at the frequency we are looking for.
4. Measure the area under the multiplied sine signal graph.
5. Repeat stages 1–4 again and again at every frequency until all the frequencies in the signal have been found.

We will not perform the whole algorithm. We don't need to. We immediately encounter a problem in stage 1. Let's set the frequency of our cosine wave to 1 to find out more.

Notice that as time increases, the area under the multiplied graph, shown in green in Figure 3, keeps growing. This is because, as t increases, the signal $x(t)$ heads for infinity. If we keep going forever, the area under the multiplied graph will grow to infinity.

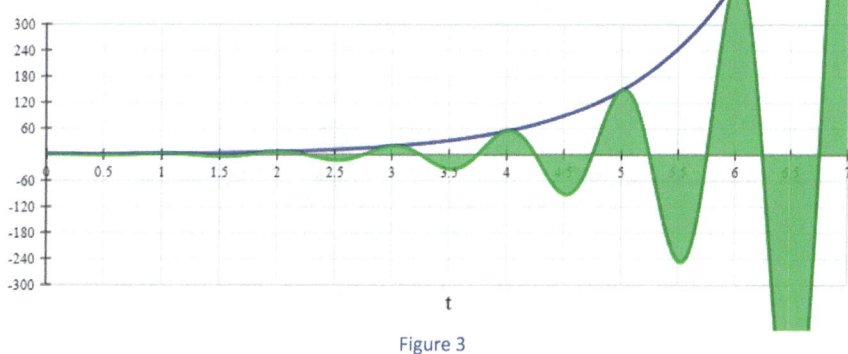

Figure 3

An integral that gives an infinite result, or a graph with an infinite area underneath it, is not something we, in our finite world, can cope with. Therefore, such signals are said to have no Fourier Transform.

So what sort of signal can the Fourier Transform model, and why is the Fourier Transform of more use to us in the real world than the Fourier Series?

The necessity of having to integrate over the whole of time means that, apart from a few theoretical exceptions, the only type of signal that can have a Fourier Transform is a signal which starts from zero amplitude when $t=-\infty$ and ends up at zero amplitude again once $t=+\infty$. Only this type of signal will have a finite area enclosed by the multiplied graph. Figure 4 shows an example of such a signal.

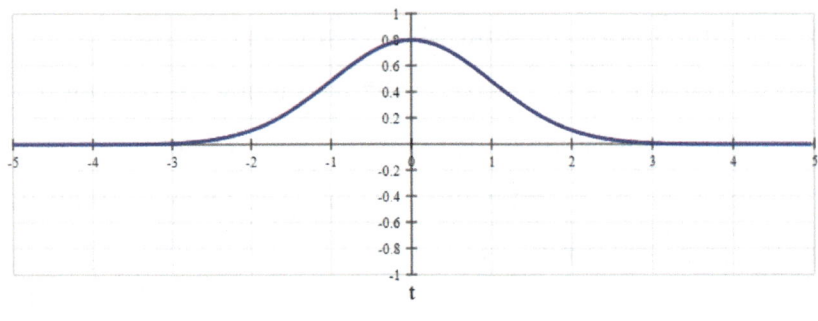

Figure 4

When we multiply the signal by the cosine wave, like in stage 1 of the Fourier Transform algorithm, and then integrate over an infinite amount of time, as shown in green in Figure 5, the result will not be infinite. The area under the graph increases only while the signal is non-zero. Once the signal returns to zero again, the value of the integral will remain constant forevermore.

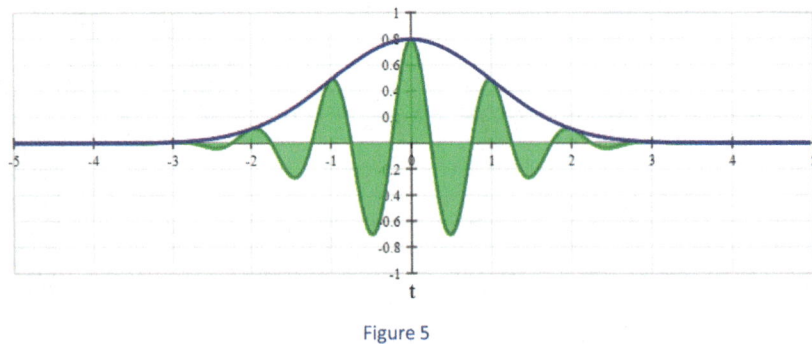

Figure 5

Luckily for us, signals which increase forever don't exist in the real world. At some point, every signal hits a maximum value and can't increase any further. All of our everyday signals have a finite magnitude and most signals exist for only a finite amount of time, meaning they all fall quiet (return to zero amplitude) eventually. Therefore, the Fourier Transform is a very useful tool for modeling real-world signals.

Continuous frequency

The second change which Dirichlet made was to turn a series of discrete results into a continuous function.

What does it mean to be discrete? Discrete, in mathematical terms, means distinct or separate. Imagine a spotlit stage like the one in Figure 6.

Figure 6

I am walking across the darkened stage, which is lit up at certain points by spotlights. When I walk into a pool of light, you can see me clearly. However, the moment I walk out of the light, you can barely see me at all. We could say, therefore, that you can only see me when I am walking through a *discrete* series of points on the stage; wherever there is a pool of light.

Now, if we were to replace the spotlights with a floodlight that lit up the whole stage, you could see me *continuously* at every point as I walked across the stage.

If we apply this analogy to Fourier's world of sine waves, the Fourier Series of a repeating signal, like the one in Figure 7, would look like Figure 8.

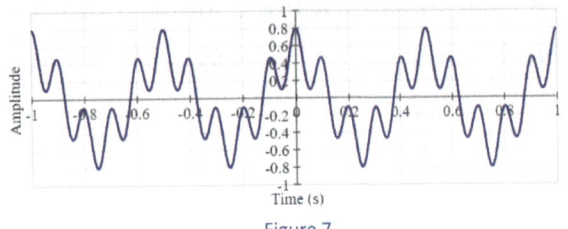
Figure 7

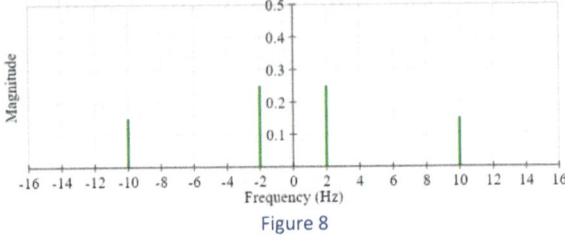
Figure 8

The output of the Fourier Series in Figure 8 is like the spotlit stage. Between the discrete frequencies in the signal, no other frequencies exist.

However, let's say we modify the signal in Figure 7, so that it fades in and fades out again, as in Figure 9. It now no longer repeats itself, and the output of its Fourier Transform looks like Figure 10.

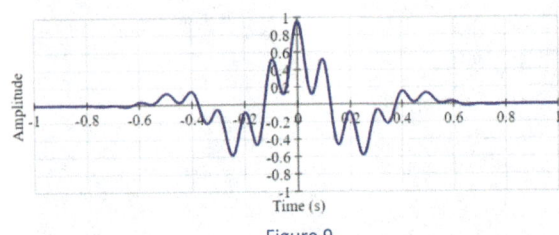
Figure 9

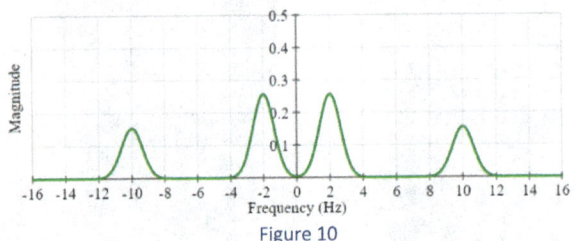
Figure 10

Around the peaks in the frequency spectrum in Figure 10, all other intermediate frequencies exist, forming a continuous function of frequencies.

Interference

Which signals have a Fourier Transform? Apart from a few theoretical exceptions, only signals that exist for a finite amount of time have a Fourier Transform.

Fourier posits that all signals are built out of a series of sinusoids. But sinusoids go on forever. If a signal built out of sinusoids is going to exist for only a finite amount of time, we somehow have to affect the magnitude of the sinusoids within the signal so that they start from zero magnitude at time $t=-\infty$ and return to zero magnitude before $t=+\infty$.

Most signals contain more than one sinusoid. If we add two sinusoids together with a similar frequency, they interfere with each other. Sometimes they interfere constructively to create a larger overall signal, and sometimes they interfere destructively and cancel each other out.

The graph in Figure 11 shows two sinusoids. The blue sinusoid has a frequency of 2 Hz and the red sinusoid has a frequency of 2.25 Hz. When we add them together, they interfere with each other, producing the signal shown in Figure 12.

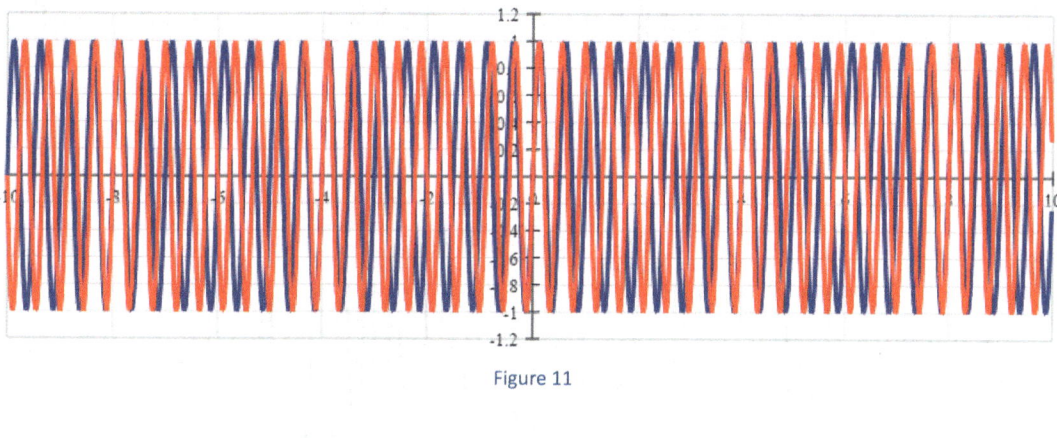

Figure 11

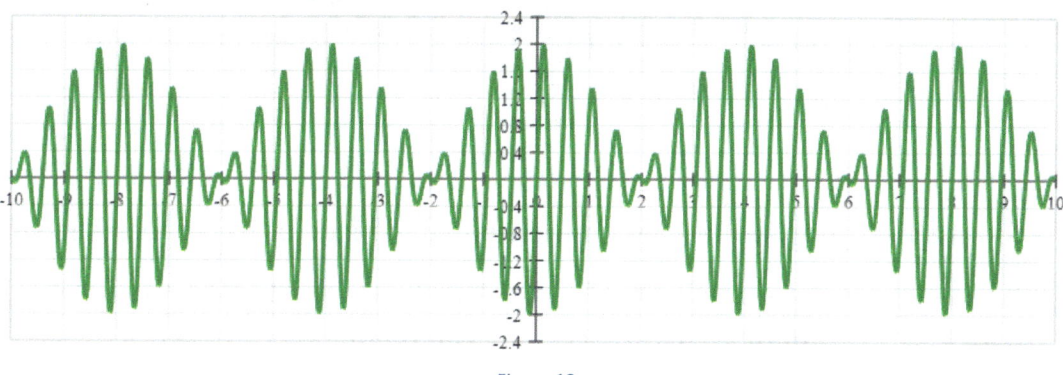

Figure 12

Beat frequencies

When the second wave from Figure 11 builds on the magnitude of the first, this is called constructive interference. When the second wave diminishes or cancels out the first, this is called destructive interference. The greater the difference between the frequency of each wave, the more times per second this constructive/destructive interference cycle occurs. This is called the beat frequency.

The beat frequency is equal to the difference between the two frequencies. So the beat frequency in Figure 12 is 0.25 Hz.

In Figure 13, we'll set the frequency of the second sinusoid to 2.05 Hz.

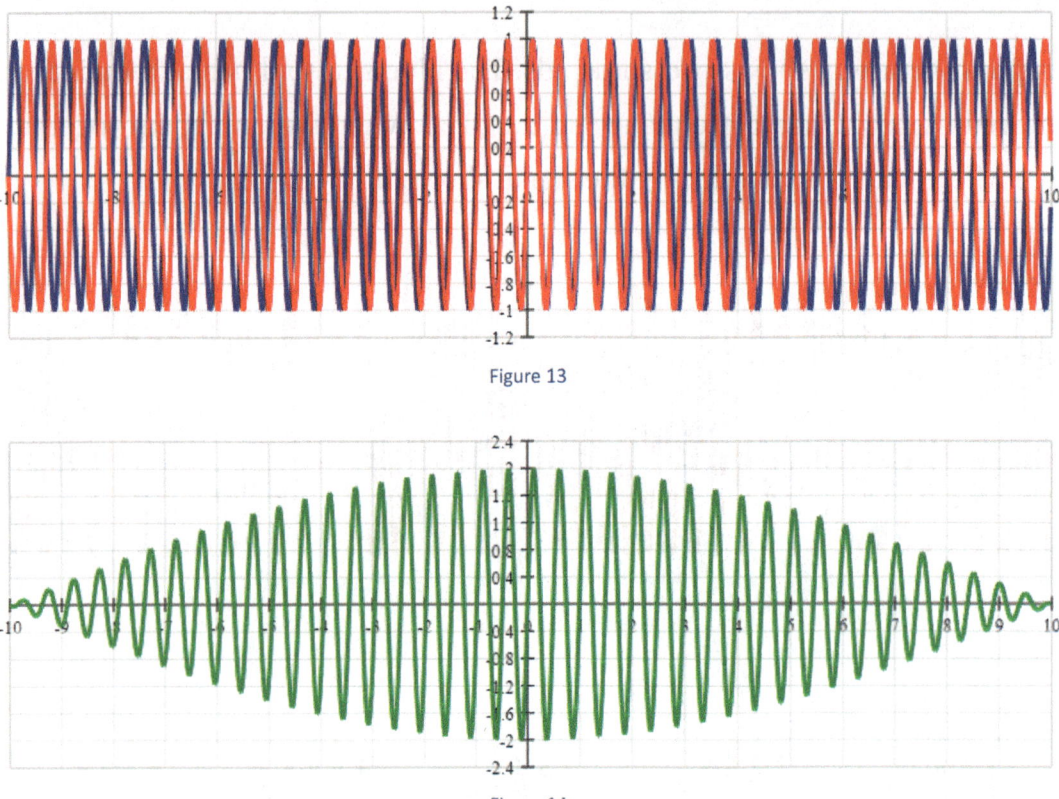

Figure 13

Figure 14

Now the signal in Figure 14 starts from zero magnitude on the left-hand side of the graph and returns to zero magnitude on the right-hand side of the graph. The shape of the graph in Figure 14 is the result of the two frequencies being very close to each other; in fact, there is only a difference of 0.05 Hz between them.

The graph in Figure 14 only shows a time range of −10 seconds to +10 seconds. If the range of the graph's x-axis was extended, we would see the signal growing again on either side of the graph. In order to return to the same situation as before, where the signal started and ended at zero in the visible part of the graph, we would have to reduce the frequency of the second wave more, until it was even closer to the first.

Imagine extending the graph so much that it would show us the signal for the whole of time, from $t=-\infty$ to $t=+\infty$. What frequency would we have to choose for the second wave so that the signal will begin and end at zero magnitude?

The nearer the frequency of the second sinusoid is to the first, the longer it takes for the second sinusoid to completely cancel out the first. If we want the second wave to cancel out the first only at $t=-\infty$ and $t=+\infty$, the frequency of the second sinusoid needs to be almost equal to the first. In fact, the two frequencies have to be so close together that the difference between them is infinitesimal.

In mathematical terms, we would say that as the frequency of the second sinusoid, f_2, approaches the frequency of the first sinusoid, f_1, the limit of the difference between the two frequencies tends to zero. We write this mathematically as shown in Equation 4.

$$\lim_{f_2 \to f_1} (f_2 - f_1) = 0$$

Equation 4

The moment this happens, we get a continuous function of frequency; there is almost no difference between two adjacent frequencies in the signal. It's like being able to see me at every point as I walk across the stage. Therefore, the graph produced by the Fourier Transform is an unbroken line.

Extending the silence

Imagine that the signal in the example above was infinite. This would mean that its magnitude began to increase (get louder) the moment time began, and won't return to zero again (fall silent) until the end of time. The signals we tend to see day-to-day are silent for much longer than that.

We know that two sinusoids, infinitely close together in frequency, can interfere destructively and cancel each other out at the beginning and end of time. But we want to be able to model much shorter signals. Let us consider the possibility of adding more sinusoids to produce more silence.

We don't have space in this book for an infinite number of sinusoids. Therefore, for this demonstration, we're going to assume that the whole of time is only 20 seconds long: 10 seconds into the past and 10 seconds into the future. We're also going to assume that the closest that two frequencies can be to each other is 0.05 Hz. We saw before that if we add a 2 Hz sinusoid to a 2.05 Hz sinusoid, both with a magnitude of 1; we get the signal shown in Figure 15.

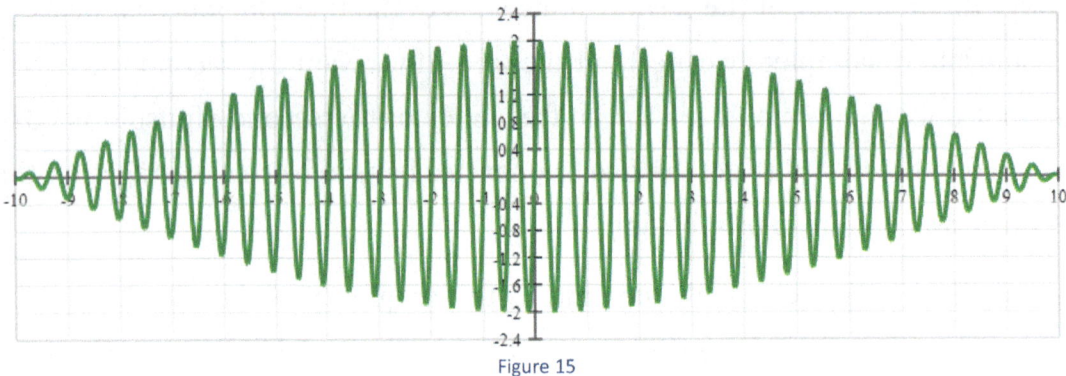
Figure 15

Let's add another sinusoid at 1.95 Hz with a magnitude of 1 and see what happens to the signal.

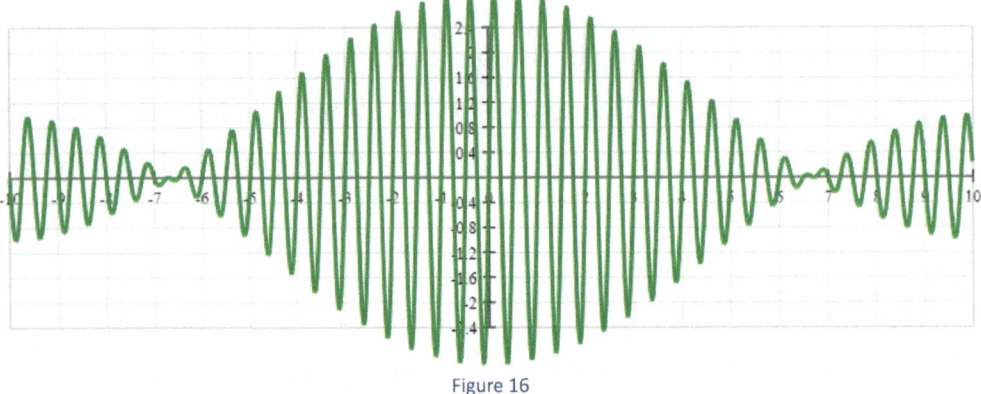
Figure 16

This cancels out the signal completely at two specific moments in time. However, it immediately grows again before and after those moments.

Let's try halving the amplitude of the 1.95 Hz and 2.05 Hz sinusoids. Now the amplitude of the original sinusoid will remain 1, but the 1.95 Hz and 2.05 Hz sinusoids will each have an amplitude of 0.5. When we add these three sinusoids together, it produces the signal shown in Figure 17.

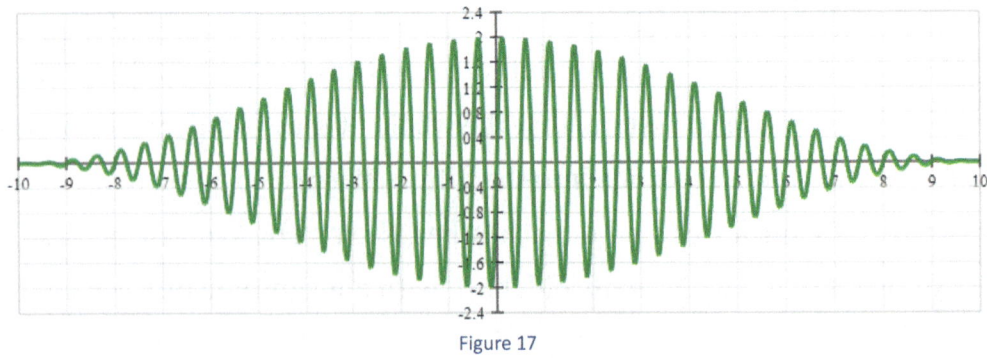
Figure 17

Now there is a little more silence at the beginning and end of the signal. Can we do even better?

Let's try adding another two sinusoids at 1.90 Hz and 2.10 Hz, both with amplitude 0.5, and raise the amplitude of the 1.95 Hz and 2.05 Hz sinusoids to 0.8.

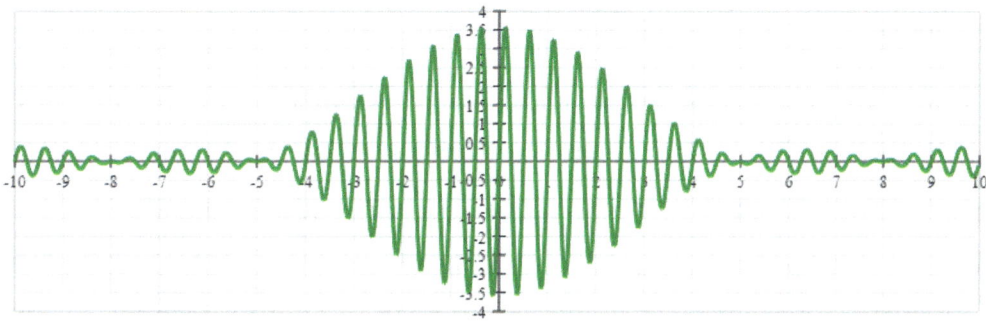

Figure 18

The signal is definitely getting shorter, but there is still some noise on either side of the signal. If we add the list of sinusoids in Table 1 together, we can make our signal silent for quite some time, as shown in Figure 19.

Frequency	Amplitude
1.75 Hz	0.00719
1.80 Hz	0.04250
1.85 Hz	0.16923
1.90 Hz	0.45404
1.95 Hz	0.82087
2.00 Hz	1.00000
2.05 Hz	0.82087
2.10 Hz	0.45404
2.15 Hz	0.16923
2.20 Hz	0.04250
2.25 Hz	0.00719

Table 1

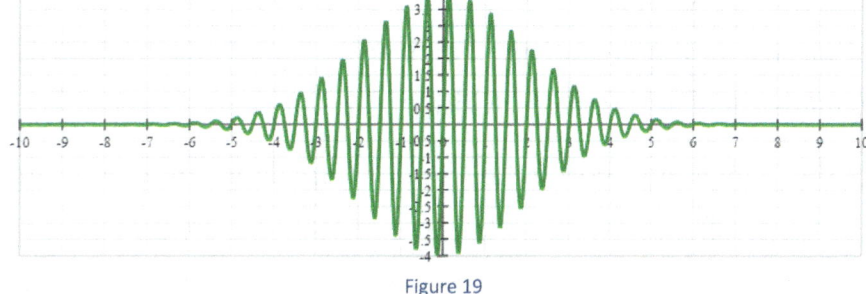

Figure 19

The values in the amplitude column in Table 1 are not random guesses. It would be very difficult to perfectly cancel out all the unwanted sinusoids at just the right moments in time by guesswork alone. These values were calculated by running a Fast Fourier Transform on a sampled version of the signal in Figure 19. We'll learn about sampling in the next chapter, and about the Fast Fourier Transform in Chapter 5. For now, the important thing to understand is that adding different sinusoids together at adjacent frequencies with just the right amplitudes causes them to cancel each other out for large portions of the signal.

Let us review. The 2 Hz sinusoid has the largest amplitude in this signal. To fade the 2 Hz sinusoid in and out, we needed additional sinusoids at adjacent frequencies. As we added sinusoids, we found that adjacent frequencies could interfere destructively with the main frequency and cancel parts of it out, but that they themselves would add to the beginning and end of the signal. So we needed additional adjacent frequencies to cause more destructive interference and cancel them out too.

Table 1 shows only 11 different frequencies. This is because we are looking at the signal for a limited period of time. If we extended the x-axis of the graph, we would see the signal repeat again and again every 20 seconds (20 seconds being the reciprocal of 0.05 Hz, the difference between the adjacent frequencies). For this signal to never repeat, the difference between adjacent frequencies would need to be infinitesimal, meaning there would be an infinite number of adjacent frequencies between 1.75 Hz and 2.25 Hz.

Bandwidth

The difference between the highest and the lowest non-zero frequencies in a signal is called the bandwidth. The band of non-zero frequencies in the signal in Figure 19 stretches from 1.75 Hz to 2.25 Hz, so it has a bandwidth of 0.5 Hz. The greater this difference is, the greater the bandwidth. There is a relationship between how long the signal is silent and the bandwidth of the signal.

As the signal gets shorter, the amount of silence on either side of the signal increases. The more silence there is, the greater the bandwidth, as we need more sinusoids at adjacent frequencies to interfere destructively with the main sinusoid and cancel it out. This canceling out generates silence. However, each new sinusoid itself adds to the signal, so we need yet more sinusoids to cancel the new one out too. This yields an increased bandwidth, as shown in Figure 20. The signal is much shorter than in the previous example, because the bandwidth is much wider, 2.5 Hz.

The frequencies present in the signal in the left-hand graph in Figure 20 are shown on the right-hand graph in Figure 20. This is called a frequency domain graph as the x-axis shows the frequency of each sinusoid in the signal while the y-axis shows the magnitude of each sinusoid.

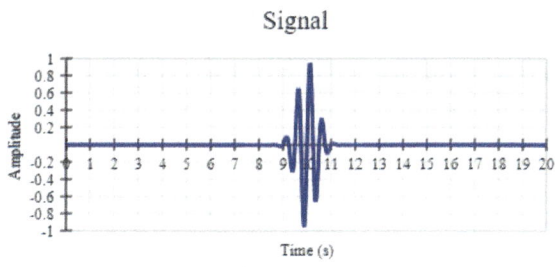
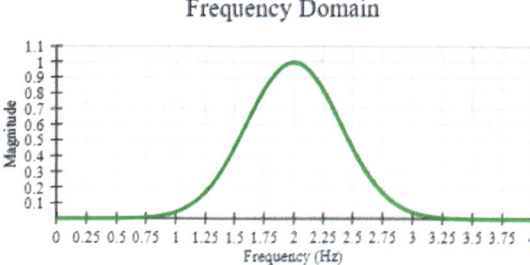

Figure 20

Conversely, as the signal gets longer, there is less silence. When there is no silence at all and the signal carries on forever, the 2.00 Hz sinusoid is the only one present. In such a case, the bandwidth is the smallest it can be; a single line showing a single sinusoid in the signal: 2.00 Hz, as shown in Figure 21.

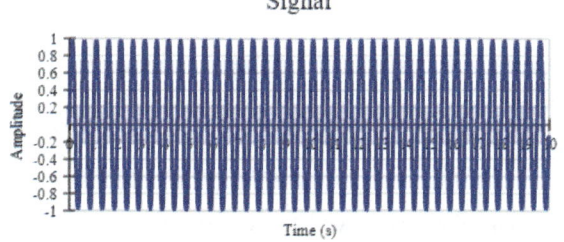
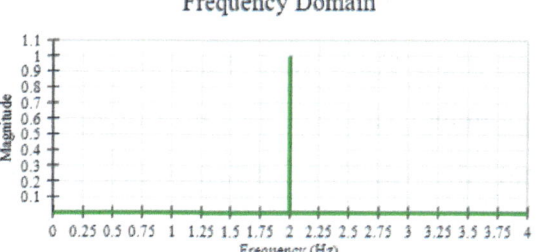

Figure 21

Coping with infinity

The changes we have examined throughout this chapter are those that Peter Gustav Lejeune Dirichlet made to the Fourier Series to turn it into the Fourier Transform. These changes are shown in green and red in Equation 5 and Equation 6.

$$c_n = \int_{-\frac{P}{2}}^{+\frac{P}{2}} x(t) \cdot e^{-i2\pi f_n t} \cdot dt \qquad X(f) = \int_{-\infty}^{+\infty} x(t) \cdot e^{-i2\pi f t} \cdot dt$$

Equation 5 – The Fourier Series Equation Equation 6 – The Fourier Transform Equation

- **Infinite Time**

    Dirichlet increased the time over which we integrate the signal, to infinity.

- **Continuous Frequency**

    Dirichlet turned a series of discrete frequencies into a continuous function, reducing the size of the jumps between adjacent frequencies until they were infinitely small.

These changes allow the Fourier Transform to model real-world signals, which don't repeat themselves and don't last forever. However, dealing with the inherent infinities involved in the Fourier Transform is difficult.

In our everyday lives, we seldom encounter signals that go on forever, so we can safely set the lower limit of the Fourier Transform integration to just before the signal started, and set the upper limit to just after the signal ended. Doing so yields exactly the same result as if we had integrated the signal over the whole of time as the Fourier Transform requires. However, if we are trying to analyze an arbitrary signal, how can we cope with the infinitesimal jumps between frequencies?

The Fourier Transform, like the Fourier Series, tests one frequency at a time and then moves on to the next. If the Fourier Transform requires us to test a series of frequencies that forms a continuous function, we'll have to test an infinite number of intermediate frequencies. This clearly isn't practical. It may be a great theoretical tool for analyzing signals described by known equations, but how can we turn the theory into something more practical and analyze an arbitrary signal whose equation we don't know?

A practical implementation of the Fourier Transform needs to be more discrete, rather than continuous. Therefore, in the next chapter, we'll look at how we can turn The Fourier Transform into the practical tool we use today, a tool that revolutionized the way we analyze and compress data. We'll begin by exploring the Discrete-Time Fourier Transform.

Chapter 2: The Discrete-Time Fourier Transform

The Discrete-Time Fourier Transform (DTFT) is a further development of the Fourier Transform. However, whereas the Fourier Transform treats time as continuous, the Discrete-Time Fourier Transform, as its name suggests, thinks of time as a discrete list of individual moments.

But why do we need the DTFT? Doesn't the Fourier Transform already do an admirable job of telling us about the frequencies that are present in everyday non-repeating signals? What is so special about the DTFT, and how does it work?

Ever since the early 1980s, discrete-time applications have become a big business, with the number of devices that make use of discrete-time growing exponentially. The following is just one example of its many applications.

A brief history of sound-recording media

In the old days, music was distributed on records or cassette tapes. Both these mediums recorded, each in its own way, a continuous analog sound signal. The instantaneous amplitude of the signal was represented, at every point in time, by some change in a property of the medium. On records, it was the shape of the groove on its surface; whereas on cassettes, it was the strength of the magnetic field stored in the ferric oxide coating on the tape.

Then, in the early 80s, the compact disc began the digital music revolution. No longer was the sound signal recorded continuously. Instead it was sampled; its amplitude was measured at individual moments in time. This meant that it was possible to store a signal as a list of numbers, encoded in a series of tiny indentations on the disc's surface. A laser in the CD player would read these indentations, and a digital circuit would then convert the numbers back into an analog sound signal.

We could therefore think of the compact disc as music's first mass-produced discrete-time storage device.

As technology improved, MP3 players became popular. People wanted to store more and more music on smaller and smaller devices. Then, music began to stream over the internet – but initially the internet was slow. It couldn't cope with multiple packets of 88,200 16-bit numbers of CD-quality audio flying through its infrastructure every second to many different people, each wanting to hear a different song.

More and more efficient methods of compression were required to compress the huge amount of data required to store and transmit the sampled sound waves.

Fourier's 200-year-old idea of representing a signal, not by its instantaneous amplitude, but by the properties of the sinusoids that make it up, was an ideal candidate for the job. It became the kernel for so many of the audio and video compression algorithms we use every single day.

The problem with the Fourier Transform

However, there was a problem. The Fourier Transform contains far too many infinities to be of practical use. Those infinities have to go!

There are three infinities inherent in the Fourier Transform.

1. The Infinite Integral – The signal is integrated over the whole of time.
2. Continuous Frequency – There are an infinite number of intermediate frequencies making up the signal.
3. Continuous Time – The signal being analyzed contains a value at every single moment in time; an infinite number of moments.

The Discrete-Time Fourier Transform didn't eliminate infinities 1 and 2, but it did eliminate infinity number 3, as its name suggests.

Comparing the Fourier Transform and the DTFT

Let's compare the equations for the Fourier Transform and the Discrete-Time Fourier Transform.

$$X(f) = \int_{-\infty}^{+\infty} x(t) \cdot e^{-i2\pi ft} \cdot dt \qquad\qquad X(f) = \sum_{n=-\infty}^{+\infty} x_n \cdot e^{-i2\pi fn}$$

Equation 7 – The Fourier Transform Equation   Equation 8 – The Discrete-Time Fourier Transform Equation

The complex exponential

Both the Fourier Transform and DTFT multiply the signal by a complex exponential, but there is a difference in the symbol used to represent time.

$$e^{-i2\pi ft} \qquad\qquad e^{-i2\pi fn}$$

Expression 1 – The Fourier Transform Complex Exponential   Expression 2 – The DTFT Complex Exponential

We replace the continuous-time variable, *t*, of the Fourier Transform with the discrete-time variable, *n*, of the DTFT. We'll meet this *n* again soon, but first let's remind ourselves of what the complex exponential is.

Euler's formula tells us that the complex exponential is a cosine wave in the real dimension and a sine wave in the imaginary dimension, as shown in Equation 9.

$$e^{-ix} = cos(x) - i \times sin(x)$$

Equation 9

When we multiply our signal by the complex exponential, we first multiply it by a cosine wave, then by an inverted sine wave. The sine wave is inverted because of the minus sign preceding the sine term. The imaginary number *i* ensures that we keep the two groups of results separate.

A discrete-time signal instead of a continuous one

Discrete time vastly reduces the infinite number of measurements required to record a signal. Instead of measuring the signal at every single point in time, we're going to sample it a finite number of times over its duration.

$$x(t) \qquad\qquad x_n$$

Expression 3 – A Continuous-Time Signal   Expression 4 – A Discrete-Time Signal

| A signal that may have looked like Figure 22 in continuous time ... | ... will now look like Figure 23 in discrete time |
|---|---|//

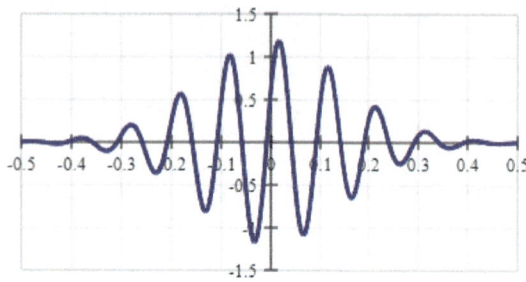

Figure 22 – A Continuous-Time Signal

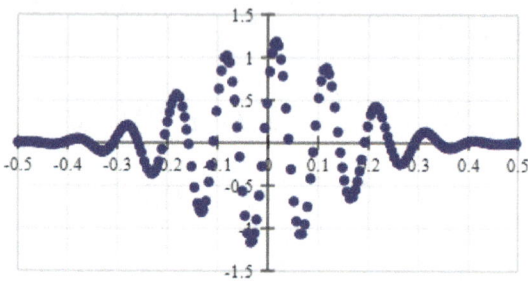

Figure 23 – A Discrete-Time Signal

We denote continuous time in the Fourier Transform by the symbol *t*, and discrete time in the DTFT by the symbol *n*, where:

$$n = -\infty, ..., -3, -2, -1, 0, 1, 2, 3, ..., +\infty$$

Summation instead of integration

The Fourier Transform equation contains an integration sign, like the one in Expression 5, whereas in the DTFT, a summation sign, like the one in Expression 6, has replaced the integration.

$$\int_{-\infty}^{\infty} (...) \cdot dt$$

Expression 5 – The Integration Sign of the Fourier Transform

$$\sum_{n=-\infty}^{\infty} (...)$$

Expression 6 – The Summation Sign of the DTFT

Integration is the act of finding the area under the graph produced by the expression inside the brackets. To do this, we cut the graph up into an infinite number of infinitely thin slices and add the area of all the slices together.

Since the DTFT operates on a discrete-time signal, it uses a summation operation to add together all the terms in the list produced by the expression inside the brackets.

Calculating the DTFT – a numerical example

To understand how the DTFT works, let's perform a DTFT on the discrete-time signal shown in Figure 24. The green points are the samples that are going to be used in the calculation.

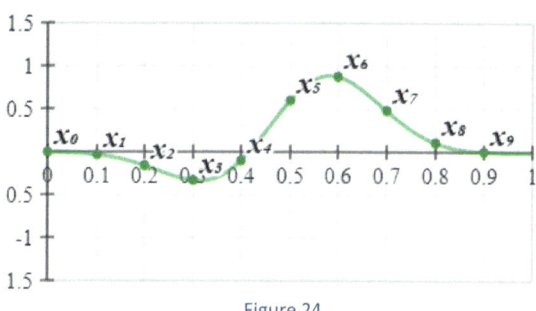

Figure 24

The signal contains 10 samples, with the values listed in Figure 25.

x_0	x_1	x_2	x_3	x_4	x_5	x_6	x_7	x_8	x_9
-0.002	-0.031	-0.158	-0.329	-0.100	0.598	0.875	0.481	0.108	0.004

Figure 25

As there are still a few infinities left in the DTFT, we'll have to take some shortcuts.

- The DTFT requires us to test an infinite number of frequencies. Our time is limited, so for the sake of this example, we'll test for only one frequency: 1 Hz.
- The DTFT looks at signals over the whole of time. As ours is limited, we'll using a signal that is zero for all but the short period contained within the graph in Figure 24.

We want to use the DTFT to answer three questions about the signal.
1. Does a 1 Hz sinusoid exist in the signal?
2. How much does it contribute to the signal (what is its magnitude)?
3. When does it occur in the signal (what is its phase)?

The first stage of performing a DTFT is to multiply all the points on the signal by the corresponding samples on a cosine wave at the test frequency. Figure 26 shows the signal and the cosine wave, and Figure 27 shows the result of the multiplication.

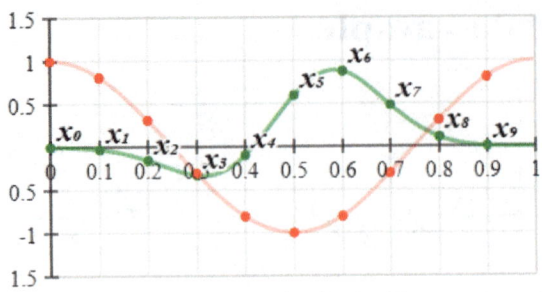

Figure 26 – Signal and Cosine Wave

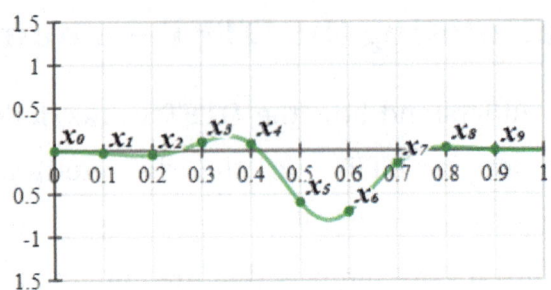

Figure 27 – Signal Multiplied by Cosine Wave

The second stage is to add all the multiplied points together to calculate the overall contribution of the cosine component at the test frequency, as shown in Figure 28.

For this signal, the contribution of the 1 Hz cosine component is –1.313.

Sample	Signal		Cosine		Result
x_0	-0.002	×	1.000	=	-0.002
x_1	-0.031	×	0.809	=	-0.025
x_2	-0.158	×	0.309	=	-0.049
x_3	-0.329	×	-0.309	=	0.102
x_4	-0.100	×	-0.809	=	0.081
x_5	0.598	×	-1.000	=	-0.598
x_6	0.875	×	-0.809	=	-0.708
x_7	0.481	×	-0.309	=	-0.149
x_8	0.108	×	0.309	=	0.033
x_9	0.004	×	0.809	=	0.003
			Sum	=	-1.313

Figure 28

We repeat stage 1 using an inverted sine wave at the test frequency.

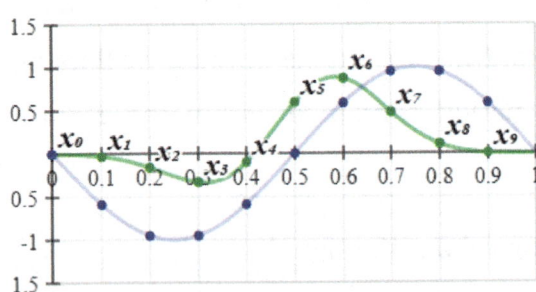

Figure 29 – Signal and Sine Wave

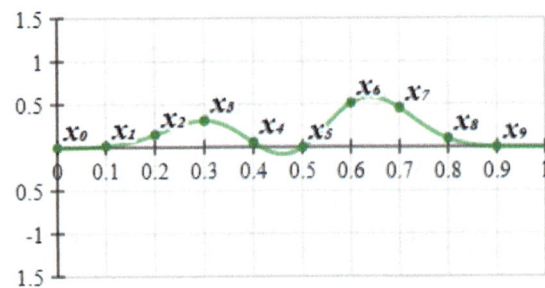

Figure 30 – Signal Multiplied by Sine Wave

Then we add all the multiplied points together, as we did in stage 2, to calculate the overall contribution of the sine component at the test frequency, as shown in Figure 31. For this signal, the contribution of the 1 Hz sine component is 1.619.

Sample	Signal		Sine		Result
x_0	-0.002	×	0.000	=	0.000
x_1	-0.031	×	-0.588	=	0.018
x_2	-0.158	×	-0.951	=	0.151
x_3	-0.329	×	-0.951	=	0.313
x_4	-0.100	×	-0.588	=	0.059
x_5	0.598	×	0.000	=	0.000
x_6	0.875	×	0.588	=	0.515
x_7	0.481	×	0.951	=	0.458
x_8	0.108	×	0.951	=	0.103
x_9	0.004	×	0.588	=	0.002
			Sum	=	1.619

Figure 31

This signal contains both a cosine and sine component at 1 Hz. The cosine component has a magnitude of −1.313 and the sine component has a magnitude of 1.619. This tells us that the signal does indeed contain a sinusoid with a frequency of 1 Hz.

To find the magnitude, M, of that sinusoid, we perform Pythagoras on the real (cosine), $\Re$, and the imaginary (sine), $\Im$, components we just calculated, as shown in Equation 10.

$$M = \sqrt{\Re^2 + \Im^2} = \sqrt{-1.313^2 + 1.619^2} = 2.084$$

Equation 10

To find the phase, P, of that sinusoid, we perform an inverse tangent on the real, $\Re$, (cosine) and the imaginary, $\Im$, (sine) components.

$$P = tan^{-1}\left(\frac{\Im}{\Re}\right) = tan^{-1}\left(\frac{1.619}{-1.313}\right) = -51°$$

Equation 11

So the DTFT provides the following answers to my questions:

1. Yes, a 1 Hz sinusoid exists in the signal.
2. It has a magnitude of 2.084.
3. It has a phase shift of -51°.

The frequency periodicity of discrete-time signals

If you've ever run the Fourier Transform on a discrete-time signal, such as the one in Figure 32, you may have noticed something strange in its frequency spectrum. As we would expect, the signal contains a group of frequencies around 10 Hz, which is the dominant frequency of the signal, as shown in Figure 33.

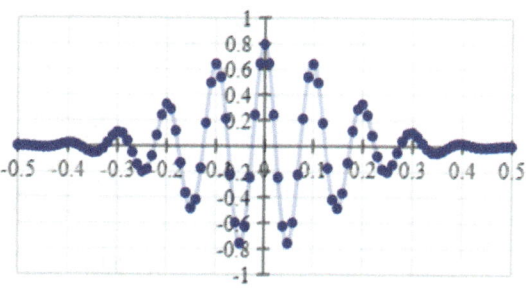

Figure 32

However, the DTFT tests an infinite number of frequencies. Around 90 Hz, and then again around 110 Hz, and yet again at 190 Hz, there are three more peaks.

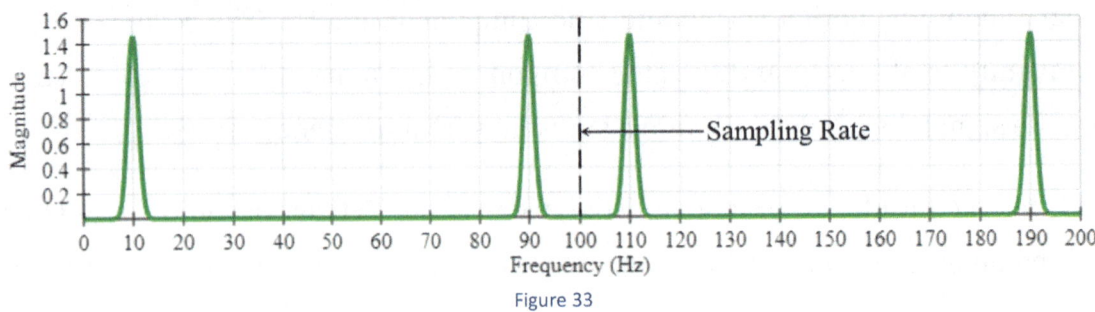

Figure 33

In fact, if we were to continue increasing the frequency, this repeating pattern of peaks would occur again and again at 100 Hz intervals forever. This particular signal was sampled 100 times every second. Therefore, its sampling rate is 100 Hz. Where do these higher frequencies come from? There is definitely no 90 Hz frequency or any higher frequency in the signal in Figure 32.

This is a consequence of the sampling process. We have thrown away some of the information we had about the signal. Both the signal and the test wave are defined only at the sampled points, represented by the blue dots in the graphs in Figure 34 to Figure 37. The DTFT can calculate a value only for those discrete points. Whatever happens between one sample and the next is invisible to the DTFT. As long as the test wave hits the same points at those moments in time for which a sample exists, then the result will be non-zero and the DTFT will think that frequency exists in the signal.

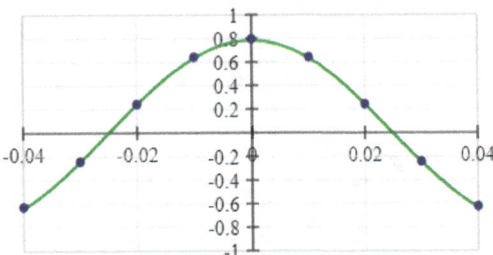

Figure 34 – Signal Samples and 10 Hz Test Wave

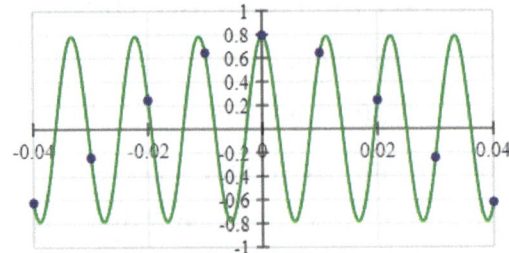

Figure 35 – Signal Samples and 90 Hz Test Wave

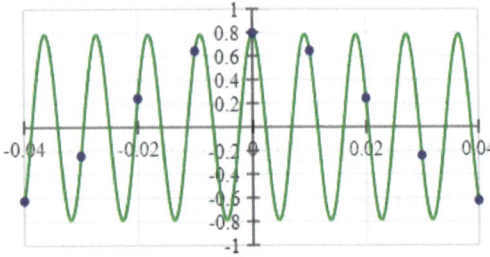

Figure 36 – Signal Samples and 110 Hz Test Wave

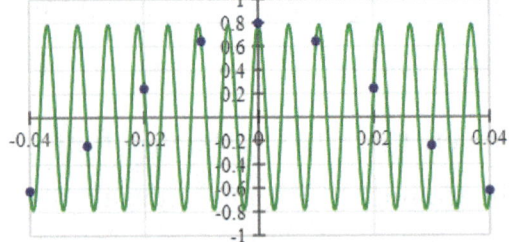

Figure 37 – Signal Samples and 190 Hz Test Wave

So even though we have a non-repeating signal in the time domain, with the DTFT, we'll get a repeating spectrum in the frequency domain. The spectrum will repeat itself every R Hz; where R is the frequency at which we sampled the signal. We sampled this signal at 100 Hz. Therefore, the rate at which the spectrum repeats itself is 100 Hz.

Aliasing and the Nyquist Rate

Now, you might think that these repeating frequencies are just a quirk of the math and that they're not actually present in the signal. However, this quirk has a very important practical ramification on the lower limit of the sampling frequency we need to choose in order to represent the signal accurately.

According to the Nyquist theorem, in order to reproduce our signal accurately out of individual samples, we need to sample it at a rate of at least twice the highest frequency present in the signal. This is called the Nyquist rate. If we sample our signal at less than the Nyquist rate, we get a phenomenon known as aliasing.

To hear an audio example of what aliasing sounds like, scan the QR code or click on the link: Aliasing

Aliasing occurs when frequencies in the signal that are higher than the Nyquist rate are reflected back into the frequency range of the original signal, distorting it. This reflection happens because there aren't enough sampled points to represent the signal accurately. You always need at least two sample points per frequency component. Therefore, you should always sample your signal at a sampling rate that is higher than the Nyquist rate.

The highest frequency present in the signal shown in Figure 38, is below 20 Hz. Therefore, according to the Nyquist theorem, If we sample the signal at 40 Hz we should be able to reproduce it accurately.

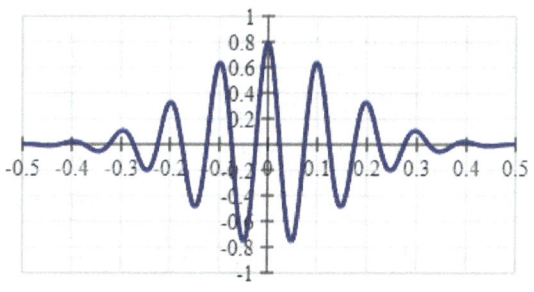

Figure 38

If we do so, then performing a DTFT on the signal gives us the graph shown in Figure 39.

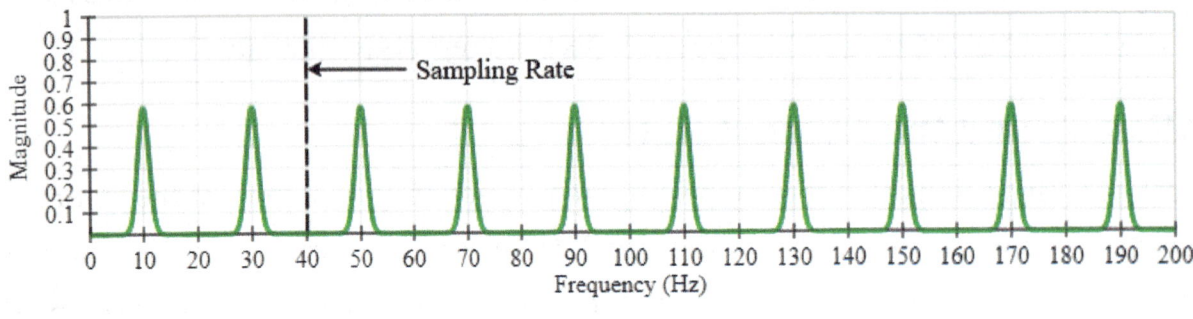

Figure 39

We can see the pattern of repeating peaks. Below the sampling rate of 40 Hz, we see one peak at 10 Hz, which we know exists in our original signal, and a second peak at 30 Hz. We know that the 30 Hz frequency doesn't really exist in the original signal but is a consequence of the sampling process. These double peaks then repeat at intervals of the sampling rate (40 Hz) as we have already seen. There is plenty of space between each peak, and the peaks are not interfering with one another.

What happens if we reduce the sampling rate to 25 Hz?

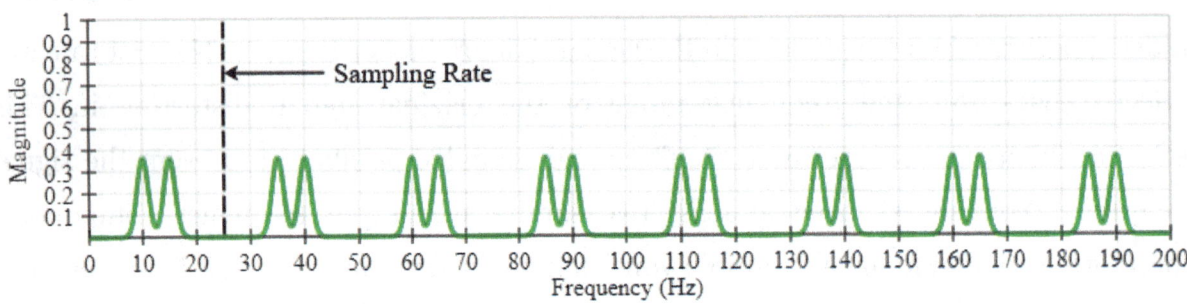

Figure 40

Now that the sampling rate in Figure 40 is only 25 Hz, the two separate peaks that were at 10 Hz and 30 Hz in Figure 39 have crashed into each other. The highest frequency in the original signal is around 14 Hz. As we are now sampling below the Nyquist rate, the higher frequencies have been reflected back into the frequency range of the original signal and have distorted it. We have lost too much information in the sampling process and we cannot now accurately reconstruct the signal.

Infinities remaining in the DTFT

The DTFT has made progress in turning the Fourier Transform into a practical tool for analyzing random signals, like sound signals, in the real world. However, there are still two more infinities to be eliminated.

- The Infinite Integral – The signal is integrated over the whole of time
- Continuous Frequency – There are an infinite number of intermediate frequencies making up the signal

In order to eliminate these infinities, we need yet another development of the Fourier Transform, The Discrete Fourier Transform, or DFT, which we will examine in the next chapter.

Chapter 3: The Discrete Fourier Transform

In the last chapter, the Discrete-Time Fourier Transform (DTFT) took the first step in turning the Fourier Transform into a practical tool. The Discrete Fourier Transform (DFT), which we'll cover in this chapter, is going to complete that journey and give us an algorithm we can actually implement on an arbitrary signal.

The problem with the Fourier Transform is that it assumes we can cope with signals that are infinite. While this is okay for certain mathematical functions with known integrals, in the real world of random signals, the idea is clearly ridiculous. We are finite beings; such powers are beyond our capabilities.

- The Fourier Transform assumes we can look at our signal over the whole of time. Clearly, we can't; no one lives forever.
- The Fourier Transform requires that we test every frequency in existence, which is an infinite number of frequencies. This is impossible; we just don't have the time.
- The Fourier Transform takes it for granted that we can measure our signal at every point in time. This is a problem, especially if we're going to use a computer to do the calculations for us.

The DTFT eliminated the third infinity by sampling the signal. Now the DFT is going to eliminate the first two infinities as well.

Comparing the DTFT and DFT

$$X(f) = \sum_{n=-\infty}^{+\infty} x_n \cdot e^{-i2\pi f n}$$

Equation 12 – The Discrete-Time Fourier Transform Equation

$$X_k = \sum_{n=0}^{N-1} x(n) \cdot e^{-i2\pi \frac{k}{N} n}$$

Equation 13 – The Discrete Fourier Transform Equation

Like the DTFT, the Discrete Fourier Transform treats time as discrete; the signal is sampled, and the discrete terms are all added together using the summation sign. However, the similarity between the two formulae ends there. In the DFT, the frequency term, *f*, and the limits of the summation, −∞ and +∞, have both been replaced by finite terms.

Testing a finite number of frequencies in the DFT

Time is discrete both in the DTFT and in the DFT. This is denoted in both formulae by *n*. The signal, *x*, is a function of discrete time, *n*, rather than of continuous time, *t*. In the DFT, however, frequency is discrete too; we test only certain specific frequencies. In the Discrete Fourier Transform, the continuous *f* of the DTFT has been replaced by the DFT's discrete *k*, where:

$$k = 0, 1, 2, 3, ..., N-1$$

In the same way that *n* is the index of the discrete-time value rather than the time that has elapsed, so too, *k* is the index of the discrete-frequency value rather than the frequency itself. Therefore, if we ever wanted to convert our frequency index, *k*, into an actual frequency value, f_k, we would need Equation 14, where:

$$f_k = R \times \frac{k}{N}$$
Equation 14

- N is the total number of times we sampled our signal.
- *k* tells us which frequency index we are testing.
- *R* is the sampling rate, or how many times per second we sampled the signal.

Similarly, if we ever wanted to convert our time index, n, into an actual measure of how much time has elapsed, t_n, we would need Equation 15.

$$t_n = \frac{n}{R}$$
Equation 15

The complex exponential

As with all forms of the Fourier Transform, the DFT multiplies the signal by a complex exponential, meaning it multiplies the signal by a cosine wave and an inverted sine wave. The sine wave is inverted because of the minus sign in the exponential. The imaginary number, *i*, ensures that we keep the cosine and sine parts of the calculation separate.

$$e^{-i2\pi f n} \qquad\qquad e^{-i2\pi \frac{k}{N} n}$$

Expression 7 – The DTFT Complex Exponential   Expression 8 – The DFT Complex Exponential

The continuous frequency term, *f*, in the DTFT's complex exponential has been replaced by the discrete frequency term, *k/N*, in the DFT's complex exponential. This is because we are now dealing with a finite number of frequencies. In the DFT, these frequencies are not absolute frequencies in Hz as they were in the DTFT, but frequencies that are a fraction of the total number of frequencies being tested, N. This is why the sampling frequency R is not present in Expression 8.

This means that instead of multiplying the signal by an infinite number of cosine and sine waves at every frequency possible, we multiply it by only a finite number of cosine and sine waves at a finite number of frequencies. The lowest frequency will be zero, and the highest frequency will be a cosine or sine wave that oscillates N—1 times over the entire sample range.

$$X(f) \qquad\qquad X_k$$

Expression 9 – A Function of Results Produced by the DTFT   Expression 10 – A Discrete List Of Results Produced by the DFT

In the same way, the continuous function, *X(f)*, of frequency terms produced by the DTFT has been replaced by a discrete list of terms, X_k, produced by the DFT.

Looking at the signal for a finite amount of time

$$\sum_{n=-\infty}^{\infty}(...) \qquad\qquad \sum_{n=0}^{N-1}(...)$$

Expression 11 – The Summation of the DTFT   Expression 12 – The Summation of the DFT

Our lives are finite; we do not live forever. Therefore, if the Fourier Transform is to be of any practical use, we cannot wait an eternity for the results. Therefore, the Discrete Fourier Transform looks at the signal for only a limited amount of time. This raises the question: how long should we look at our signal? Ideally, we would look at the whole of the signal from start to finish. But what happens if we don't know when the signal is going to end? Or maybe we want results quickly, or even in real time, and don't have time to wait for the whole signal to finish.

One way to tackle this is to divide the signal into blocks of equal size and deal with each block separately. We run the Discrete Fourier Transform on one block at a time and get a result that is specific to that block.

What factors do we have to consider when deciding on how big a block to use?

Many factors may be important in this decision; for example, processing time and memory, to name but two. However, we will concentrate on one factor in particular.

Frequency resolution in the DFT

Perhaps the biggest consideration when running the DFT is frequency resolution. The major difference between the DTFT and the DFT is the fact that now, not only is time discrete, but the frequency is too. We are giving up on some of the frequency information available in the signal. In the time domain, we had to ensure that we sampled at or above the Nyquist rate. Now, in the frequency domain, we must run our DFT on enough of the signal to ensure that we capture all the frequencies that interest us. Frequency resolution is the ability to distinguish between two or more closely spaced frequencies in a signal. The size of the step between two adjacent frequencies tells us the frequency resolution; the smaller the step, the higher the resolution. Figure 41 shows a frequency-domain graph with high frequency resolution, and Figure 42 shows a frequency domain graph with low frequency resolution.

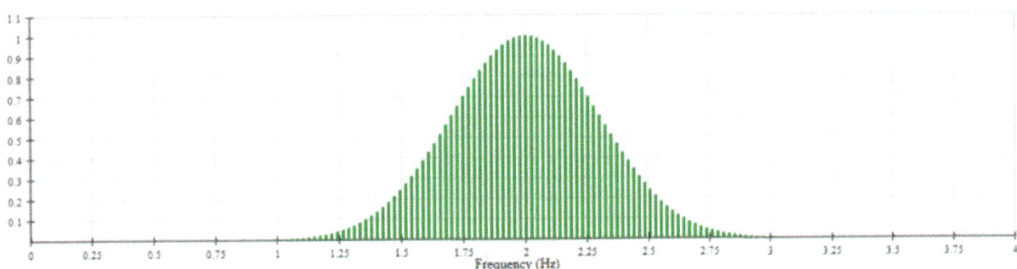

Figure 41 – High Frequency Resolution / Small Steps between Frequencies

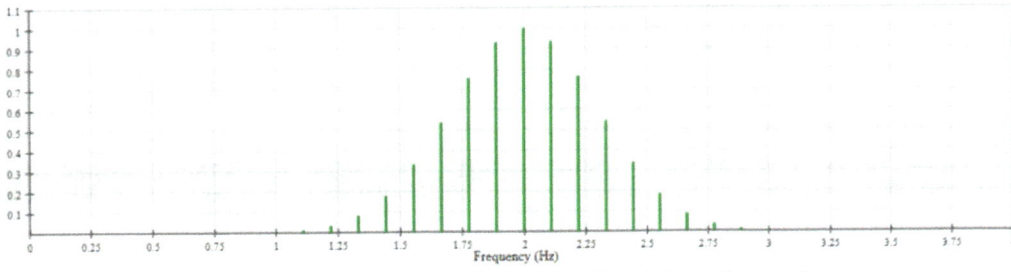

Figure 42 – Low Frequency Resolution / Large Steps between Frequencies

Calculating the frequency step size

Frequency resolution, and its reciprocal, the size of the step between two adjacent frequencies, is a function of two factors:

- The sampling rate, *R*; how many times we sampled the signal each second.
- The number of samples, *N*; how many times we sampled our signal overall.

If we want to know how well our DFT is going to be able to distinguish between adjacent frequencies in the signal, we need to know the frequency step size, *Δf*, which we calculate using Equation 16.

$$\Delta f = \frac{R}{N}$$

Equation 16

So, for a given sampling rate, *R*, to improve the frequency resolution of our DFT, we need to make the step size, *Δf*, smaller by increasing the number of samples, N, in our signal; in other words, we need to record the signal for longer.

Calculating the DFT – a numerical example

Now that the Discrete Fourier Transform has finally given us a practical way of performing a Fourier Transform on a real signal, let's do an example.

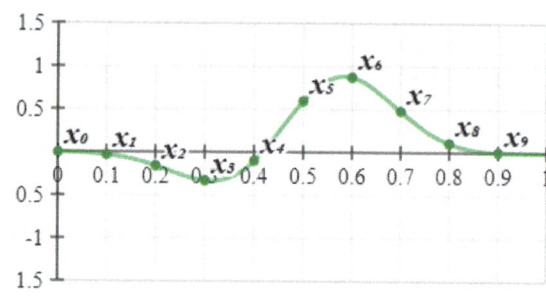

We're going to calculate the DFT for the signal in Figure 43, the same signal we used in the numerical example for the DTFT in the last chapter.

Figure 43

However, as the DFT has eliminated all the remaining infinities of the Fourier Transform, we can calculate all the frequencies present in the signal, rather than just a single frequency like we did in the DTFT.

The signal contains ten samples, as listed in Figure 44.

x_0	x_1	x_2	x_3	x_4	x_5	x_6	x_7	x_8	x_9
0.00	-0.03	-0.16	-0.33	-0.10	0.60	0.88	0.48	0.11	0.00

Figure 44

Using Euler's formula, we convert the exponential form of the DFT equation, Equation 17, into its polar form, Equation 18.

$$X_k = \sum_{n=0}^{N-1} x_n \cdot e^{-i2\pi \frac{k}{N} n}$$

$$X_k = \sum_{n=0}^{N-1} x_n \cdot \left[\cos\left(2\pi \frac{k}{N} n\right) - i \cdot \sin\left(2\pi \frac{k}{N} n\right) \right]$$

Equation 17 – The DFT in Its Exponential Form

Equation 18 – The DFT in Its Polar Form

We need the DFT in its polar form because we are going to calculate the cosine and sine components for each frequency index, *k*.

We can understand Equation 18 as a list of instructions:

1. Multiply each sample in the signal, x_n, by $\cos(2\pi \frac{k}{N} n)$ at frequency index *k*=0, producing a multiplied cosine wave.
2. Add all the points on the multiplied cosine wave together, producing the cosine component for frequency index *k*=0.
3. Multiply each sample in the signal, x_n, by $-\sin(2\pi \frac{k}{N} n)$ at frequency index *k*=0, producing a multiplied sine wave.
4. Add all the points on the multiplied sine wave together, producing the sine component for frequency index *k*=0.
5. Repeat steps 1–4, incrementing *k* each time until *k* = *N*–1.

There are ten samples in the signal, which means there will be ten frequency indices to test. In the DFT the number of frequencies tested is always equal to the number of samples in the part of the signal being analyzed by the DFT. So in Equation 18, *n* will range from 0 to 9, and *k* will also range from 0 to 9.

The result for each frequency index, X_k, will be a complex number whose real part will be equal to the cosine contribution and whose imaginary part will be equal to the sine contribution.

For each frequency index, *k*, I have drawn two graphs, the first showing the signal overlaid with a cosine wave and the second showing the signal overlaid with an inverted sine wave. Below each graph is a table listing the individual sample values taking part in each calculation.

Calculating the DFT of frequency index k=0

Cosine Component

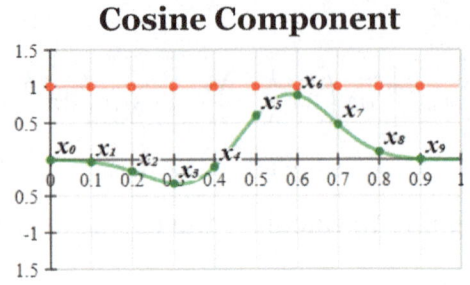

Figure 45

Sine Component

Figure 46

Sample	Signal		Cosine		Result
x_0	-0.002	×	1	=	-0.002
x_1	-0.031	×	1	=	-0.031
x_2	-0.158	×	1	=	-0.158
x_3	-0.329	×	1	=	-0.329
x_4	-0.1	×	1	=	-0.1
x_5	0.598	×	1	=	0.598
x_6	0.875	×	1	=	0.875
x_7	0.481	×	1	=	0.481
x_8	0.108	×	1	=	0.108
x_9	0.004	×	1	=	0.004
			Sum	=	1.45

Figure 47

Sample	Signal		Sine		Result
x_0	-0.002	×	0	=	0
x_1	-0.031	×	0	=	0
x_2	-0.158	×	0	=	0
x_3	-0.329	×	0	=	0
x_4	-0.1	×	0	=	0
x_5	0.598	×	0	=	0
x_6	0.875	×	0	=	0
x_7	0.481	×	0	=	0
x_8	0.108	×	0	=	0
x_9	0.004	×	0	=	0
			Sum	=	0

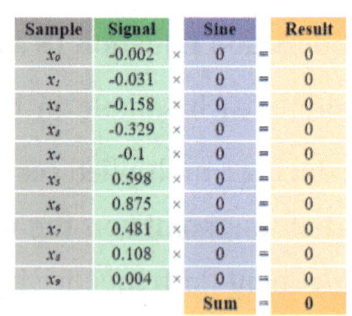

Figure 48

$$X_{k=0} = 1.45 + 0i$$

Calculating the DFT of Frequency index k=1

Cosine Component

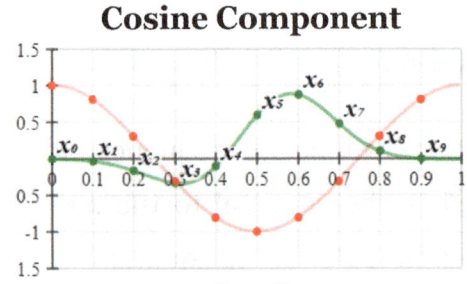

Figure 49

Sine Component

Figure 50

Sample	Signal		Cosine		Result
x_0	-0.002	×	1	=	-0.002
x_1	-0.031	×	0.809	=	-0.025
x_2	-0.158	×	0.309	=	-0.049
x_3	-0.329	×	-0.309	=	0.102
x_4	-0.1	×	-0.809	=	0.081
x_5	0.598	×	-1	=	-0.598
x_6	0.875	×	-0.809	=	-0.708
x_7	0.481	×	-0.309	=	-0.149
x_8	0.108	×	0.309	=	0.033
x_9	0.004	×	0.809	=	0.003
			Sum	=	-1.31

Figure 51

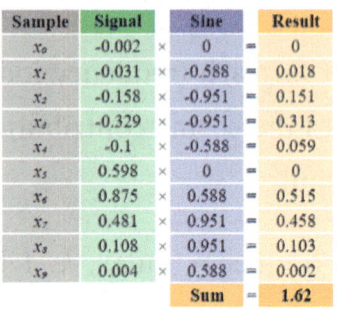

Sample	Signal		Sine		Result
x_0	-0.002	×	0	=	0
x_1	-0.031	×	-0.588	=	0.018
x_2	-0.158	×	-0.951	=	0.151
x_3	-0.329	×	-0.951	=	0.313
x_4	-0.1	×	-0.588	=	0.059
x_5	0.598	×	0	=	0
x_6	0.875	×	0.588	=	0.515
x_7	0.481	×	0.951	=	0.458
x_8	0.108	×	0.951	=	0.103
x_9	0.004	×	0.588	=	0.002
			Sum	=	1.62

Figure 52

$$X_{k=1} = -1.31 + 1.62i$$

Calculating the DFT of frequency index $k=2$

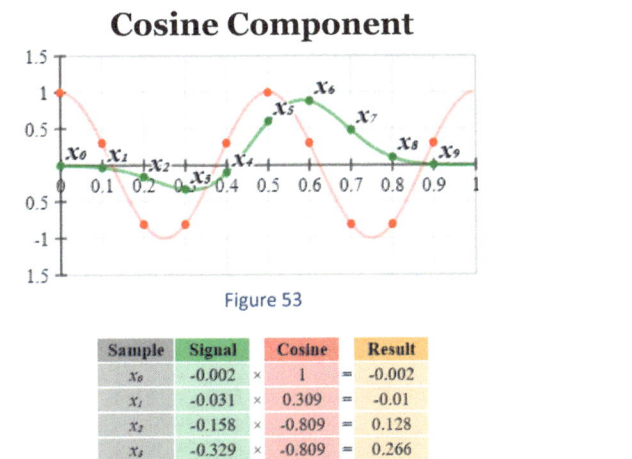

Cosine Component

Figure 53

Sine Component

Figure 54

Sample	Signal		Cosine		Result
x_0	-0.002	×	1	=	-0.002
x_1	-0.031	×	0.309	=	-0.01
x_2	-0.158	×	-0.809	=	0.128
x_3	-0.329	×	-0.809	=	0.266
x_4	-0.1	×	0.309	=	-0.031
x_5	0.598	×	1	=	0.598
x_6	0.875	×	0.309	=	0.271
x_7	0.481	×	-0.809	=	-0.389
x_8	0.108	×	-0.809	=	-0.088
x_9	0.004	×	0.309	=	0.001
			Sum	=	**0.74**

Figure 55

Sample	Signal		Sine		Result
x_0	-0.002	×	0	=	0
x_1	-0.031	×	-0.951	=	0.03
x_2	-0.158	×	-0.588	=	0.093
x_3	-0.329	×	0.588	=	-0.194
x_4	-0.1	×	0.951	=	-0.095
x_5	0.598	×	0	=	0
x_6	0.875	×	-0.951	=	-0.833
x_7	0.481	×	-0.588	=	-0.283
x_8	0.108	×	0.588	=	0.064
x_9	0.004	×	0.951	=	0.003
			Sum	=	**-1.21**

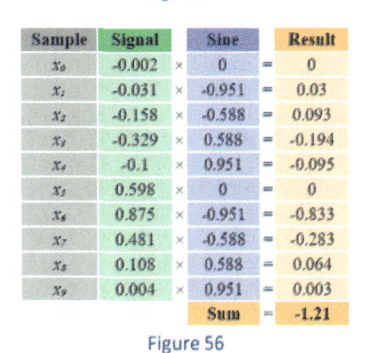

Figure 56

$$X_{k=2} = 0.74 - 1.21i$$

Calculating the DFT of frequency index $k=3$

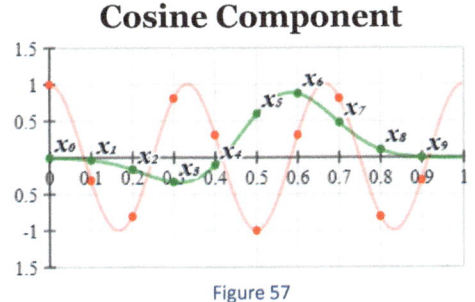

Cosine Component

Figure 57

Sine Component

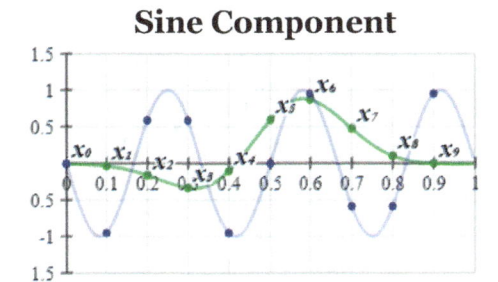

Figure 58

Sample	Signal		Cosine		Result
x_0	-0.002	×	1	=	-0.002
x_1	-0.031	×	-0.309	=	0.01
x_2	-0.158	×	-0.809	=	0.128
x_3	-0.329	×	0.809	=	-0.266
x_4	-0.1	×	0.309	=	-0.031
x_5	0.598	×	-1	=	-0.598
x_6	0.875	×	0.309	=	0.271
x_7	0.481	×	0.809	=	0.389
x_8	0.108	×	-0.809	=	-0.088
x_9	0.004	×	-0.309	=	-0.001
			Sum	=	**-0.19**

Figure 59

Sample	Signal		Sine		Result
x_0	-0.002	×	0	=	0
x_1	-0.031	×	-0.951	=	0.03
x_2	-0.158	×	0.588	=	-0.093
x_3	-0.329	×	0.588	=	-0.194
x_4	-0.1	×	-0.951	=	0.095
x_5	0.598	×	0	=	0
x_6	0.875	×	0.951	=	0.833
x_7	0.481	×	-0.588	=	-0.283
x_8	0.108	×	-0.588	=	-0.064
x_9	0.004	×	0.951	=	0.003
			Sum	=	**0.33**

Figure 60

$$X_{k=3} = -0.19 + 0.33i$$

Calculating the DFT of frequency index *k*=4

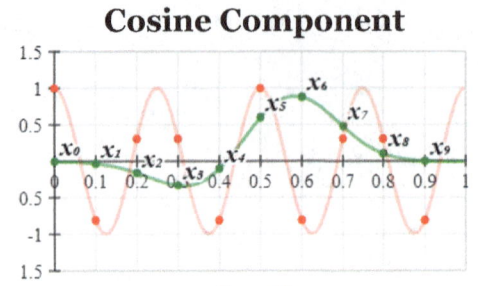

Cosine Component

Figure 61

Sine Component

Figure 62

Sample	Signal		Cosine		Result
x_0	-0.002	×	1	=	-0.002
x_1	-0.031	×	-0.809	=	0.025
x_2	-0.158	×	0.309	=	-0.049
x_3	-0.329	×	0.309	=	-0.102
x_4	-0.1	×	-0.809	=	0.081
x_5	0.598	×	1	=	0.598
x_6	0.875	×	-0.809	=	-0.708
x_7	0.481	×	0.309	=	0.149
x_8	0.108	×	0.309	=	0.033
x_9	0.004	×	-0.809	=	-0.003
			Sum	=	**0.02**

Figure 63

Sample	Signal		Sine		Result
x_0	-0.002	×	0	=	0
x_1	-0.031	×	-0.588	=	0.018
x_2	-0.158	×	0.951	=	-0.151
x_3	-0.329	×	-0.951	=	0.313
x_4	-0.1	×	0.588	=	-0.059
x_5	0.598	×	0	=	0
x_6	0.875	×	-0.588	=	-0.515
x_7	0.481	×	0.951	=	0.458
x_8	0.108	×	-0.951	=	-0.103
x_9	0.004	×	0.588	=	0.002
			Sum	=	**-0.04**

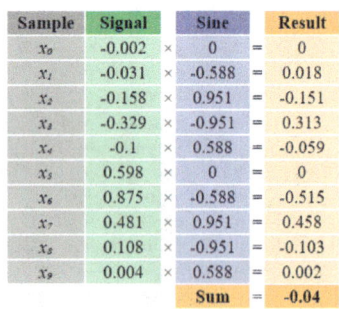

Figure 64

$$X_{k=4} = 0.02 - 0.04i$$

Calculating the DFT of frequency index *k*=5

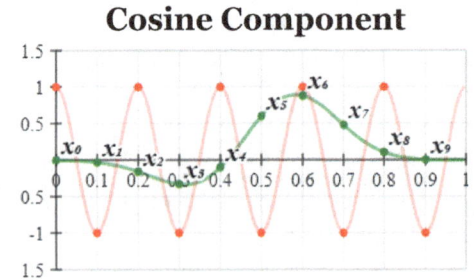

Cosine Component

Figure 65

Sine Component

Figure 66

Sample	Signal		Cosine		Result
x_0	-0.002	×	1	=	-0.002
x_1	-0.031	×	-1	=	0.031
x_2	-0.158	×	1	=	-0.158
x_3	-0.329	×	-1	=	0.329
x_4	-0.1	×	1	=	-0.1
x_5	0.598	×	-1	=	-0.598
x_6	0.875	×	1	=	0.875
x_7	0.481	×	-1	=	-0.481
x_8	0.108	×	1	=	0.108
x_9	0.004	×	-1	=	-0.004
			Sum	=	**0**

Figure 67

Sample	Signal		Sine		Result
x_0	-0.002	×	0	=	0
x_1	-0.031	×	0	=	0
x_2	-0.158	×	0	=	0
x_3	-0.329	×	0	=	0
x_4	-0.1	×	0	=	0
x_5	0.598	×	0	=	0
x_6	0.875	×	0	=	0
x_7	0.481	×	0	=	0
x_8	0.108	×	0	=	0
x_9	0.004	×	0	=	0
			Sum	=	**0**

Figure 68

$$X_{k=5} = 0 + 0i$$

Calculating the DFT of frequency index k=6

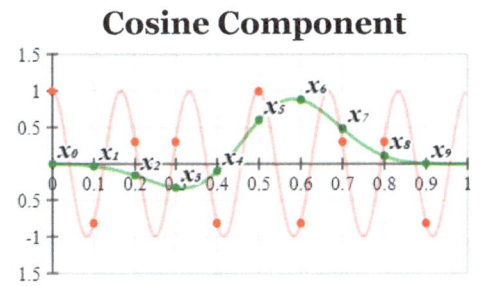

Cosine Component

Figure 69

Sine Component

Figure 70

Sample	Signal		Cosine		Result
x_0	-0.002	×	1	=	-0.002
x_1	-0.031	×	-0.809	=	0.025
x_2	-0.158	×	0.309	=	-0.049
x_3	-0.329	×	0.309	=	-0.102
x_4	-0.1	×	-0.809	=	0.081
x_5	0.598	×	1	=	0.598
x_6	0.875	×	-0.809	=	-0.708
x_7	0.481	×	0.309	=	0.149
x_8	0.108	×	0.309	=	0.033
x_9	0.004	×	-0.809	=	-0.003
			Sum	=	0.02

Figure 71

Sample	Signal		Sine		Result
x_0	-0.002	×	0	=	0
x_1	-0.031	×	0.588	=	-0.018
x_2	-0.158	×	-0.951	=	0.151
x_3	-0.329	×	0.951	=	-0.313
x_4	-0.1	×	-0.588	=	0.059
x_5	0.598	×	0	=	0
x_6	0.875	×	0.588	=	0.515
x_7	0.481	×	-0.951	=	-0.458
x_8	0.108	×	0.951	=	0.103
x_9	0.004	×	-0.588	=	-0.002
			Sum	=	0.04

Figure 72

$$X_{k=6} = 0.02 + 0.04i$$

Calculating the DFT of frequency index k=7

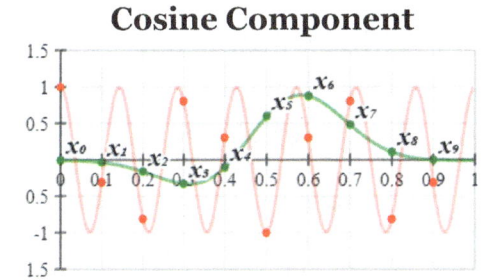

Cosine Component

Figure 73

Sine Component

Figure 74

Sample	Signal		Cosine		Result
x_0	-0.002	×	1	=	-0.002
x_1	-0.031	×	-0.309	=	0.01
x_2	-0.158	×	-0.809	=	0.128
x_3	-0.329	×	0.809	=	-0.266
x_4	-0.1	×	0.309	=	-0.031
x_5	0.598	×	-1	=	-0.598
x_6	0.875	×	0.309	=	0.271
x_7	0.481	×	0.809	=	0.389
x_8	0.108	×	-0.809	=	-0.088
x_9	0.004	×	-0.309	=	-0.001
			Sum	=	-0.19

Figure 75

Sample	Signal		Sine		Result
x_0	-0.002	×	0	=	0
x_1	-0.031	×	0.951	=	-0.03
x_2	-0.158	×	-0.588	=	0.093
x_3	-0.329	×	-0.588	=	0.194
x_4	-0.1	×	0.951	=	-0.095
x_5	0.598	×	0	=	0
x_6	0.875	×	-0.951	=	-0.833
x_7	0.481	×	0.588	=	0.283
x_8	0.108	×	0.588	=	0.064
x_9	0.004	×	-0.951	=	-0.003
			Sum	=	-0.33

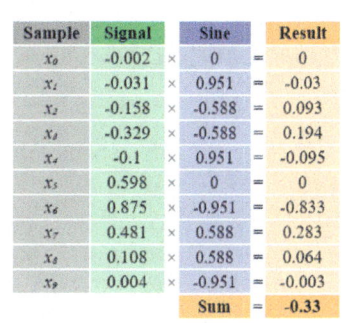

Figure 76

$$X_{k=7} = -0.19 - 0.33i$$

Calculating the DFT of frequency index *k*=8

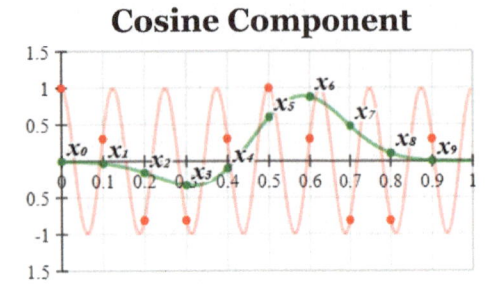

Cosine Component

Figure 77

Sine Component

Figure 78

Sample	Signal		Cosine		Result
x_0	-0.002	×	1	=	-0.002
x_1	-0.031	×	0.309	=	-0.01
x_2	-0.158	×	-0.809	=	0.128
x_3	-0.329	×	-0.809	=	0.266
x_4	-0.1	×	0.309	=	-0.031
x_5	0.598	×	1	=	0.598
x_6	0.875	×	0.309	=	0.271
x_7	0.481	×	-0.809	=	-0.389
x_8	0.108	×	-0.809	=	-0.088
x_9	0.004	×	0.309	=	0.001
			Sum	=	**0.74**

Figure 79

Sample	Signal		Sine		Result
x_0	-0.002	×	0	=	0
x_1	-0.031	×	0.951	=	-0.03
x_2	-0.158	×	0.588	=	-0.093
x_3	-0.329	×	-0.588	=	0.194
x_4	-0.1	×	-0.951	=	0.095
x_5	0.598	×	0	=	0
x_6	0.875	×	0.951	=	0.833
x_7	0.481	×	0.588	=	0.283
x_8	0.108	×	-0.588	=	-0.064
x_9	0.004	×	-0.951	=	-0.003
			Sum	=	**1.21**

Figure 80

$$X_{k=8} = 0.74 + 1.21i$$

Calculating the DFT of frequency index *k*=9

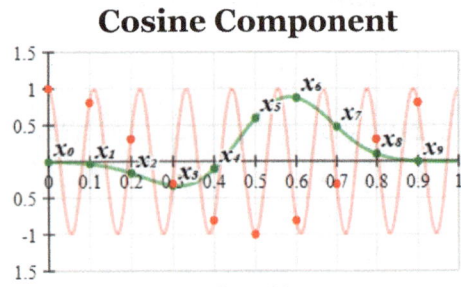

Cosine Component

Figure 81

Sine Component

Figure 82

Sample	Signal		Cosine		Result
x_0	-0.002	×	1	=	-0.002
x_1	-0.031	×	0.809	=	-0.025
x_2	-0.158	×	0.309	=	-0.049
x_3	-0.329	×	-0.309	=	0.102
x_4	-0.1	×	-0.809	=	0.081
x_5	0.598	×	-1	=	-0.598
x_6	0.875	×	-0.809	=	-0.708
x_7	0.481	×	-0.309	=	-0.149
x_8	0.108	×	0.309	=	0.033
x_9	0.004	×	0.809	=	0.003
			Sum	=	**-1.31**

Figure 83

Sample	Signal		Sine		Result
x_0	-0.002	×	0	=	0
x_1	-0.031	×	0.588	=	-0.018
x_2	-0.158	×	0.951	=	-0.151
x_3	-0.329	×	0.951	=	-0.313
x_4	-0.1	×	0.588	=	-0.059
x_5	0.598	×	0	=	0
x_6	0.875	×	-0.588	=	-0.515
x_7	0.481	×	-0.951	=	-0.458
x_8	0.108	×	-0.951	=	-0.103
x_9	0.004	×	-0.588	=	-0.002
			Sum	=	**-1.62**

Figure 84

$$X_{k=9} = -1.31 - 1.62i$$

The DFT for our signal

All the above calculations leave us with the list of results presented in Table 3.

Frequency Index	Real	Imaginary
0	1.45	0
1	-1.31	1.62
2	0.74	-1.21
3	-0.19	0.33
4	0.02	-0.04
5	0	0
6	0.02	0.04
7	-0.19	-0.33
8	0.74	1.21
9	-1.31	-1.62

Table 2

Using Equation 19, which we first met on page 28, we can calculate the frequency, f_k, from the frequency index, k. The signal was sampled 10 times in 1 second, so R, the sampling rate, is 10 Hz and there are 10 samples in the signal so N is 10.

$$f_k = R \times \frac{k}{N}$$
Equation 19

Using Equation 20, which we first met on page 22, we can calculate the magnitude from the real and imaginary parts of each result.

$$M = \sqrt{\Re^2 + \Im^2}$$
Equation 20

Using Equation 21, which we first met on page 22, we can calculate the phase from the real and imaginary parts of each result.

$$P = tan^{-1}\left(\frac{\Im}{\Re}\right)$$
Equation 21

Frequency Index	Frequency	Magnitude	Phase
0	0 Hz	1.45	0°
1	1 Hz	2.08	-51°
2	2 Hz	1.42	-59°
3	3 Hz	0.38	-60°
4	4 Hz	0.04	-63°
5	5 Hz	0.00	0°
6	6 Hz	0.04	63°
7	7 Hz	0.38	60°
8	8 Hz	1.42	59°
9	9 Hz	2.08	51°

Table 3

We can then plot these values on a magnitude graph, Figure 85, and a phase graph, Figure 86, to get a visual representation of the DFT for this signal.

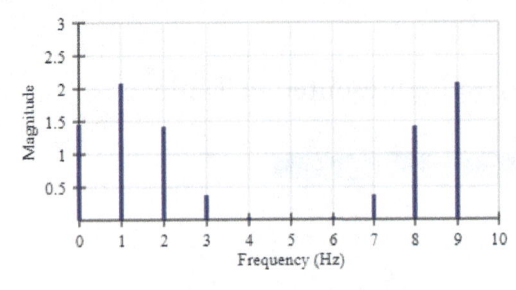

Figure 85 – Magnitude Graph

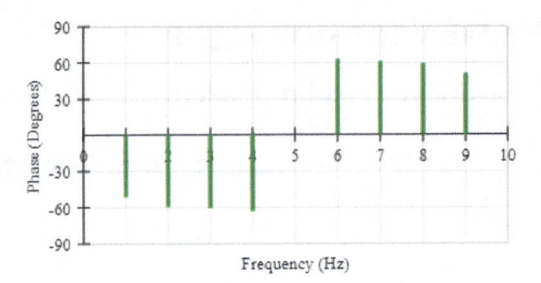

Figure 86 – Phase Graph

Long signals and short DFTs

We have now used the Discrete Fourier Transform to extract the frequency, magnitude, and phase information for a real sampled signal. However, we are still left with a problem. This signal fits nicely into the block size I have chosen. It is zero at the beginning of the block, and returns to zero before the end of the block. However, not all signals are that convenient. Most of the time, because of practical limitations, we are forced to chop a long signal into smaller blocks.

What would happen if the signal were not zero at the beginning or end of the block?

This would lead to a problem known as spectral leakage. We cannot solve this problem completely, but we can reduce it by using a method known as windowing.

In the next chapter, we'll look at what spectral leakage is and how windowing helps to reduce it.

Chapter 4: Windowing Functions

In expanding the capabilities of the Fourier Series, Dirichlet introduced two infinities to the Fourier Transform:

1. The infinite integral – We integrate the signal over the whole of time.
2. Continuous frequency – There are an infinite number of adjacent frequencies making up the signal.

Over the last few chapters, we've been systematically eliminating all those infinities so that we can run the Fourier Transform on a real-world, arbitrary signal. You may well have noticed that we've basically undone all the changes which Dirichlet made and ended up with something that is not too far away from the Fourier Series we began with.

As we learned in Chapter 1 of this book, the Fourier Series can model only repeating signals. This is because it uses the repeating cosine and sine functions to build those signals. So how can we now use what is basically a discrete time version of the Fourier Series to model non-repeating signals?

We know that our signal does not go on forever; and we know how much of that signal we fed into the Discrete Fourier Transform. We also know that if we reconstruct the signal from its frequency information, that signal will repeat itself, as we found out when we learned about Beat frequencies in Chapter 1. Therefore, we simply disregard the repeat.

However, this raises the question of what we do if we have a very long signal and don't have the time or processing power to run a DFT on the whole signal at once. The answer is that we break it down into smaller blocks and run the DFT one block at a time.

The problem is, cutting up a signal into blocks can have an adverse effect on its frequency spectrum.

Why do we need windowing functions?

One common application of the Fourier Transform is in speech recognition algorithms. Imagine you were dictating a letter into your phone. The phone recognizes what you are saying by analyzing the pattern of frequencies in your voice as you speak. Each sound has its own distinct frequency pattern. The phone analyzes each pattern and correlates it with the words in a dictionary it has in its memory.

When dictating, you want to speak naturally. You definitely don't want to enunciate each sound separately, ensuring plenty of silence between sounds to help the computer identify each individual sound.

Therefore, the computer records your voice continually. However, it constantly breaks this recording into blocks and sends each block off for processing as you speak.

Ideally, each individual syllable of a word should occupy exactly one block. However, until it analyzes the recording, your phone does not know where each syllable begins and ends. Also, computers use the Fast Fourier Transform (FFT) to analyze the data. In the next chapter we'll cover the FFT, which is a more efficient version of the Discrete Fourier Transform. One limitation of the FFT though, is that the number of samples, N, in a block needs to be a power of two. For example, N could be 2^8, which is 256 samples.

When analyzing the frequency patterns of your speech, your phone needs to maintain a uniform frequency resolution between blocks. It must also ensure that the frequencies it is analyzing in your speech are the same frequencies that exist in its word dictionary. This way it can compare the two and decide which word you are saying. As we discovered in the last chapter, for any given sample rate, the frequency resolution is determined by the number of samples in each block. Therefore, to maintain a constant frequency resolution between blocks, your phone must ensure that the size of each block remains the same. For example, it could decide that there are 256 samples in each block which would be a suitable length for the FFT.

What happens if you are in the middle of a word when the block suddenly ends? The computer simply continues analyzing the frequencies in the next block of data and builds up a model of how the frequencies in your voice change as you speak.

Windowing functions smooth the signal between blocks, ensuring that it fades in at the beginning of a block and fades out again before the end of the block. They also ensure that a phenomenon known as spectral leakage does not occur.

What is spectral leakage?

Let's consider a simple signal made of three cosine waves added together. Each cosine wave is generated using Expression 13.

$$A \times cos(2\pi ft - P)$$
Expression 13

Table 4 lists the amplitude, A, the frequency, f, and the phase, P for each wave.

Wave	A	f	P
1	1	1 Hz	100°
2	0.3	3 Hz	30°
3	0.8	4 Hz	60°

Table 4

When added together, these three waves produce the signal shown in Figure 87.

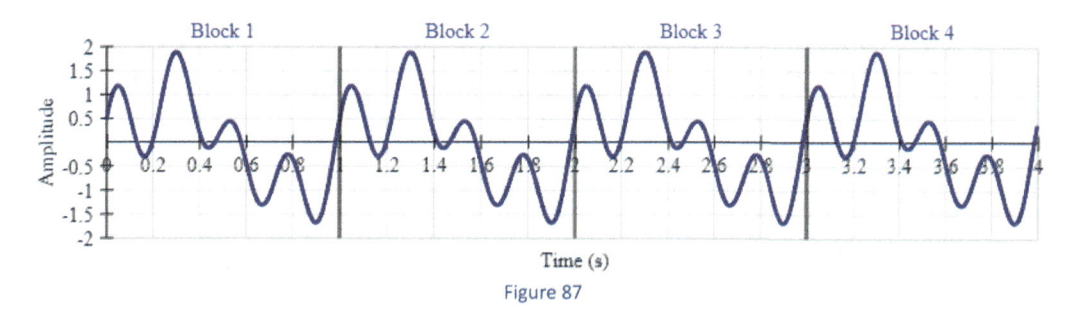

Figure 87

Let us divide the signal into four blocks of one second each. Right away we notice that this signal isn't zero at the beginning or end of the block. Does that mean we're going to need a windowing function before we run the DFT? Figure 88 shows the DFT for the first block of the signal.

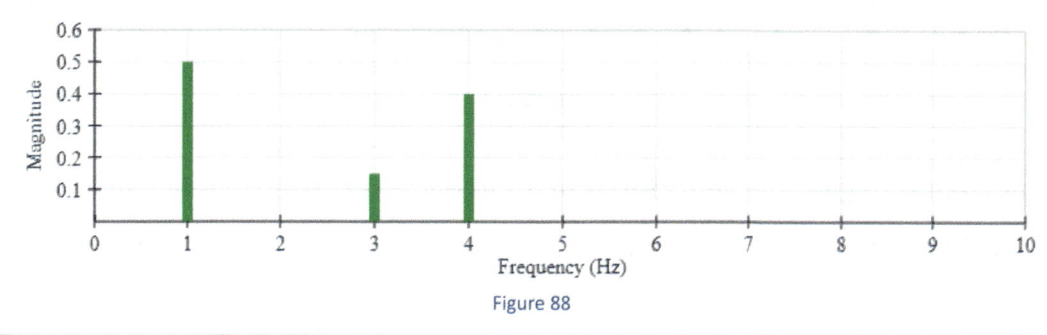

Figure 88

The three sinusoids making up the signal show up clearly in the frequency domain graph in Figure 88. There are no other sinusoids at any other frequency. We can see them clearly at 1 Hz, 3 Hz, and 4 Hz. There has been no spectral leakage, and consequently no windowing function is required.

Note: The graph zooms in on the three frequencies that interest us. Remember, as we mentioned in Chapter 2, there will be an additional three frequencies mirrored about the Nyquist rate. We can't see them on this graph, because the x-axis of the graph doesn't extend far enough.

Changing the block size

What would happen if our block size were a little shorter?

In Figure 89, we divide the same signal into five blocks. Therefore, each block contains less of the signal.

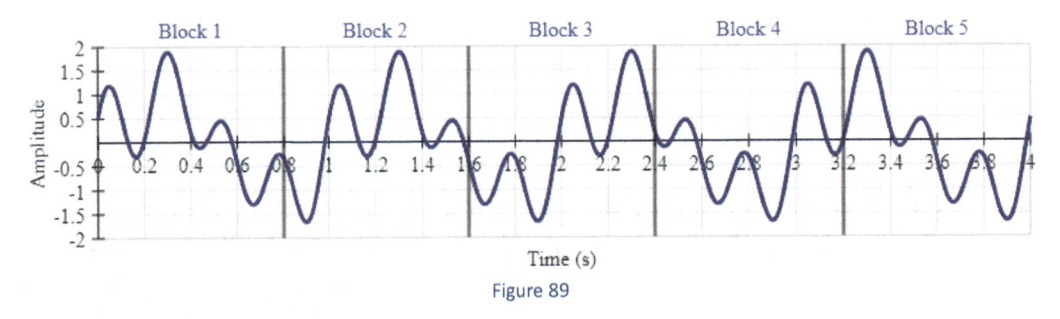

Figure 89

When we run the DFT on the first block, look at what happens to the frequency domain graph in Figure 90.

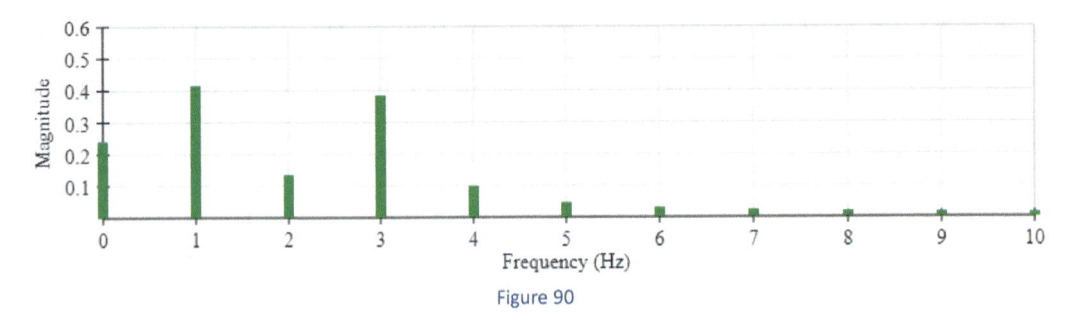

Figure 90

The same signal displays a very different frequency spectrum. How can this be? We haven't changed the signal itself; we just shortened the length of each block.

Changing the block size appears to have caused some of the energy from the three frequency components that we saw in Figure 88 to leak into the adjacent frequencies. This is called spectral leakage.

Why has changing the block size had this effect on the frequency spectrum of the signal, and how will windowing functions help us reduce the problem?

The signal in Figure 87 and Figure 89 is a repeating signal. The DFT is like a Discrete Fourier Series. Therefore, it can really model only repeating signals. Whereas in Figure 87 we ended the block exactly at the point where the signal repeats, in Figure 89, the signal has not completed its cycle by the time the block ends. If we were to reconstruct the signal from the frequency spectrum in Figure 90, it would look like Figure 91. Notice the discontinuities between each block. Rather than flowing smoothly, the amplitude of the signal suddenly jumps as the waveform repeats prematurely.

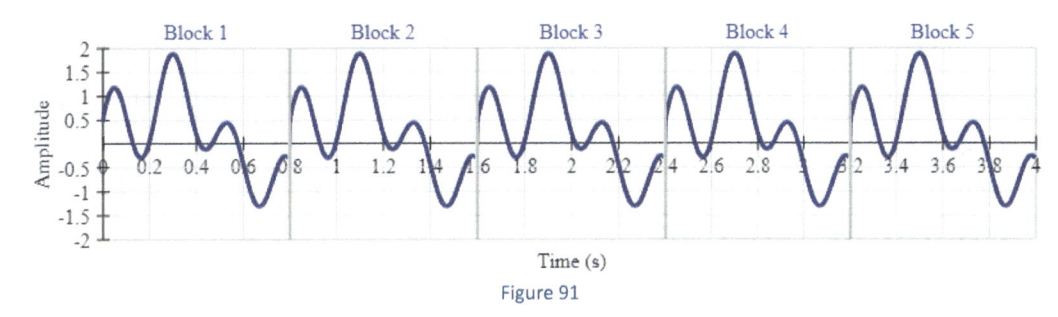

Figure 91

I carefully constructed the original signal using sinusoids at specific frequencies. When chopped into four blocks, all the sinusoids present in the signal finish their cycles at the point where the block ends. When chopped into five blocks, all the blocks are too small to contain a whole number of complete cycles, thus causing spectral leakage. Everyday signals comprise many sinusoids at many frequencies, all finishing their cycles at different times. There is no single point where we can chop up the signal, without one or more of its sinusoids still being in the middle of a cycle.

This causes a discontinuity at the beginning and end of each block. The sudden jump in amplitude introduces frequency artifacts that do not exist in the original signal. If we could smooth out this discontinuity, we could go some way to reducing the problem by removing these artifacts. This is what windowing functions do.

Different windowing functions

By definition, windowing functions change the shape of the signal, affecting its frequency spectrum. In many applications, this is preferable to the frequency artifacts caused by chopping up the signal. It is unfortunately the price we'll have to pay if we don't have the time or processing power to analyze the whole signal in one enormous block.

There are many windowing functions, but they all have one common aim: to ensure that the signal is zero, or almost zero, at the beginning and end of the block. Much like peering through a window into your home reveals only a portion of the room; windowing functions reveal only a portion of the signal. To apply the window, each point of the time domain signal is multiplied by a corresponding window coefficient. Therefore, if there are N samples of the signal in each block, there will also be N coefficients in the windowing function. Let's look at a few examples.

What follows is not an exhaustive list of every windowing function in existence; rather, it provides a taste of a few different window types.

Hann windowing function

The Hann windowing function is named after Julius Ferdinand von Hann (1839–1921). He invented a weighted moving-average technique for combining meteorological data from neighboring regions. Blackman and Tukey later used his technique to derive the Hann function, shown in Figure 92. Figure 93 shows two signals. The green signal is the original signal, while the blue signal has had three Hann windows applied to it, one after the other. If we were dividing the signal into blocks, we would split it at the point where each window ends and the amplitude is zero.

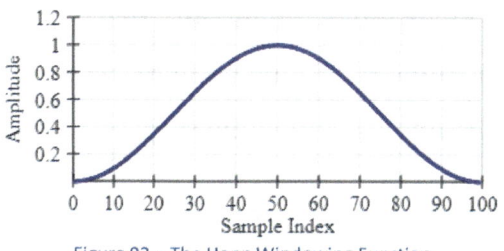

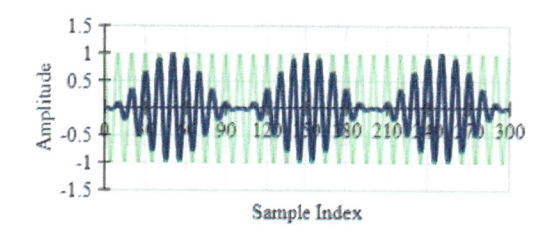

Figure 92 – The Hann Windowing Function

Figure 93 – A Signal with the Hann Window Applied

$$w_n = 0.5 - 0.5 \cdot \cos\left(\frac{2\pi n}{N}\right)$$

Equation 22 – The Hann Window Equation

Hamming windowing function

The Hamming windowing function was invented by Richard Hamming (1915–1998). It is similar in shape to the Hann window, but there is one important difference. Whereas the Hann window fades out the signal completely, the Hamming window leaves a little of the signal at the beginning and end of the block. We'll discuss why this could be useful shortly.

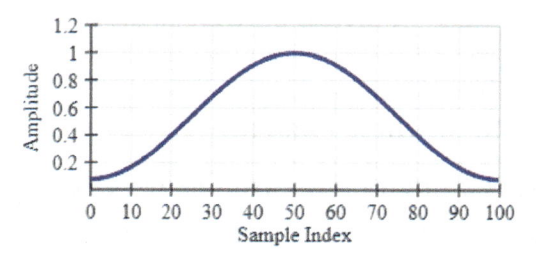

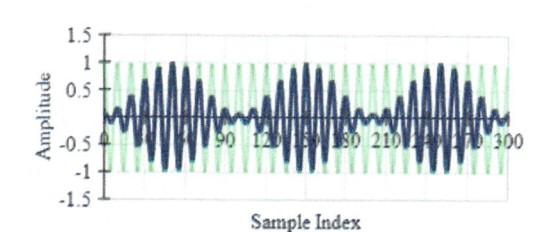

Figure 94 – The Hamming Windowing Function

Figure 95 – A Signal with the Hamming Window Applied

$$w_n = 0.54 - 0.46 \cdot \cos\left(\frac{2\pi n}{N}\right)$$

Equation 23 – The Hamming Window Equation

The combination of the 0.54 offset and the 0.46 scaling term ensures that the windowing function never touches zero. Why might we not want zero amplitude at the beginning and end of our block?

The Hamming function is employed in scenarios where we require a precise representation of the signal in the frequency domain, free from unwanted frequency artifacts. At the same time, we must retain the ability to accurately reconstruct a time domain signal after some modification performed in the frequency domain.

For example, imagine we have two audio recordings made in two different environments, and we want to try and make them sound as if they were recorded in the same environment. We could achieve this by performing the following steps:

- Copy the first signal. Keep the original version of the first signal for later.
- Apply a Hamming function to the copy of the first signal to smooth the transition between blocks and reduce the spectral leakage as much as possible.
- Convert the windowed signal from the time domain to the frequency domain using the DFT.
- Analyze the frequency spectrum of the first signal. Now we know its frequency spectrum, we have finished with the first signal.
- Apply a Hamming function to the second signal to smooth the transition between blocks and reduce the spectral leakage as much as possible.
- Convert the windowed signal from the time domain to the frequency domain using the DFT.
- Manipulate the frequency spectrum of the second signal so that it more closely matches the frequency spectrum of the first.
- Convert the modified second signal back into the time domain.
- Apply an inverse Hamming function to eliminate the windowing effect and recover the second signal.

We should now have two signals that exhibit similar frequency spectra and therefore sound as if they were recorded in similar environments.

We couldn't recover the second signal with a Hann function, because a Hann function cannot be reversed. When we apply a Hann window, by multiplying the samples by the Hann function, the first and last samples in each block are zero. We could never reverse the Hann process and recover the original value of the signal, since any value times zero is zero.

Therefore, to remedy this, the Hamming function doesn't fade the signal out completely. There is still some signal at the beginning and end of each block that can be rescaled back to its original value. There is a cost, however: the signal is not zero at the beginning and end of the block, and therefore some small discontinuities remain between blocks. As a result, the Hamming window doesn't reduce spectral leakage quite as well as the Hann window.

Bartlett windowing function

The Bartlett windowing function, invented by the English statistician Maurice Stevenson Bartlett (1910–2002), produces a triangular window.

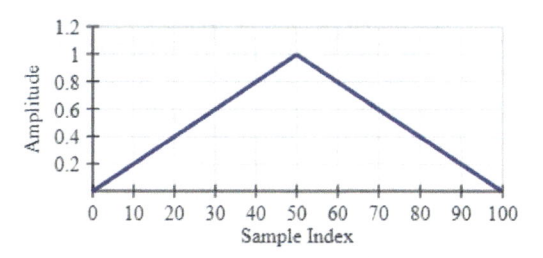

Figure 96 – The Bartlett Windowing Function

Figure 97 – A Signal with the Bartlett Window Applied

$$w(n) = \begin{cases} \frac{2n}{N} & 0 \leq n \leq \frac{N}{2} \\ 2 - \frac{2n}{N} & \frac{N}{2} \leq n \leq N \end{cases}$$

Equation 24 – The Bartlett Window Equation

Like the Hann window, it is zero at the beginning and end of the block. The linear nature of the window tapers a signal without generating too much ripple in the frequency domain.

Tukey windowing function

The Tukey windowing function is named after its inventor: John Wilder Tukey (1915–2000), whom we mentioned above as one of the developers of the Hann window. He also invented, together with James Cooley, the Cooley-Tukey FFT algorithm, which we'll examine in greater detail in the next chapter.

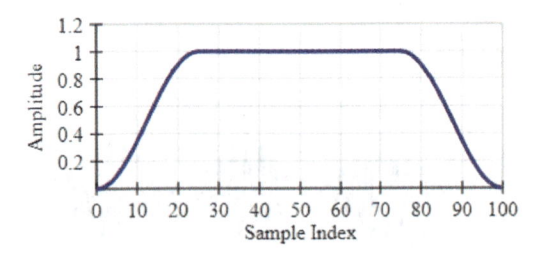

Figure 98 – The Tukey Windowing Function

Figure 99 – A Signal with the Tukey Window Applied

$$w(n) = \begin{cases} 0.5 - 0.5 \times cos\left(2\pi \frac{n}{\alpha N}\right) & 0 \leq n \leq \frac{\alpha N}{2} \\ 1 & \frac{\alpha N}{2} \leq n \leq N - \frac{\alpha N}{2} \\ 0.5 - 0.5 \times cos\left(2\pi \frac{N-n}{\alpha N}\right) & N - \frac{\alpha N}{2} \leq n \leq N \end{cases}$$

Equation 25 – The Tukey Window Equation

Notice the α in the equation. We can assign α any value from 0 to 1. When α = 0, the Tukey window resembles a rectangular function, and when α = 1, the Tukey window resembles a Hann function.

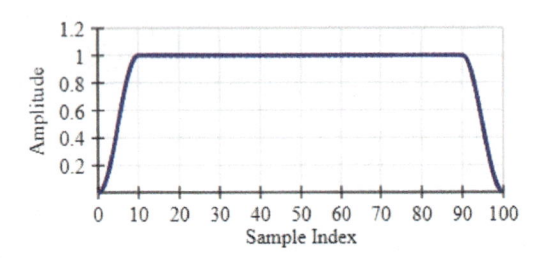

Figure 100 – The Tukey Windowing Function: α=0.2

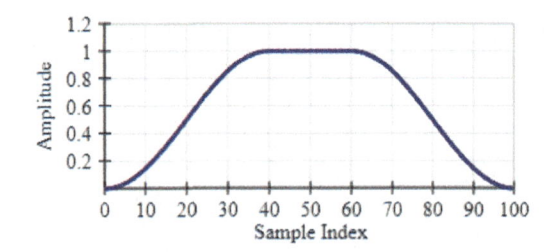

Figure 101 – The Tukey Window Function: α=0.8

Choosing a windowing function

The question of which is the best windowing function to use on your signal is complicated. It depends on what you intend to do with your signal once you've used the DFT to convert it into the frequency domain. For example, if you want to recover the original time domain signal, you might choose a Hamming window. Selecting the windowing function is therefore a bit of an art.

For an interactive simulation comparing how the different windowing functions affect the frequency-domain representation of the signal in Figure 87, scan the QR code or click on the following link:

Windowing Function Demo

Notice particularly how all windowing functions reduce the spectral leakage into the higher frequencies, as they were designed to do.

We have seen that windowing functions help condition a signal before we break it up into the blocks required by the DFT. They remove the discontinuities between blocks which lead to spectral leakage, the leaking of some of the signal's energy into higher frequencies. These discontinuities are artifacts generated by the way the DFT sees the signal, and they arise because of the DFT's similarity to the Fourier Series, which can model only repeating signals.

Using the DFT and windowing functions, at last, we've found a way of performing a Fourier Transform on real signals. However, the DFT can be very processor hungry, especially if there are many samples in each block. To be of any practical use, the DFT needs to be much more efficient.

In the next chapter, we'll discover how a researcher from IBM and a professor from Princeton University transformed the Fourier Transform from a tool that only highly skilled mathematicians could use, into one of the most important and fundamental signal and data analysis tools in use today. Known as the Cooley-Tukey algorithm, you may well have heard of it under its more popular name: The Fast Fourier Transform.

Chapter 5: The Fast Fourier Transform

The Fourier Transform is everywhere. Almost every day you pick up a piece of technology which implements it. However, this wouldn't be the case if a way hadn't been found to make it easier to calculate.

The problem lies in the sheer number of computations required for even a few samples. For example, in order to calculate the Discrete Fourier Transform for a signal containing a mere 256 samples, those 256 samples must first be multiplied by 256 points on a cosine wave and added together, then by 256 points on a sine wave and added together. This entire process then needs to be repeated with cosine and sine waves at 256 different frequencies. We need to perform 65,536 complex multiply-and-add operations for only 256 samples.

When the story of what we now know as the Fast Fourier Transform began, computers were in their infancy, processors were slow, and computational efficiency was the name of the game.

Enter James Cooley and John Tukey. These two American mathematicians published a paper in 1965 in which they proposed a recursive algorithm that vastly reduced the number of operations required. This algorithm became known as the Cooley-Tukey Fast Fourier Transform. In the 1960s, the invention of the silicon transistor was about to revolutionize the world of computing. Cooley and Tukey's work came at just the right time and their names became forever associated with the FFT.

Repeating calculations

The secret of the Fast Fourier Transform's efficiency is the number of samples in each block. If this number is a power of 2, then the calculations repeat themselves. Remembering the result of the calculation saves having to repeat the same calculation again. Also, if the samples are ordered in a certain way, the result from one set of calculations can form the starting point of the next.

How the Fourier Transform works

In order to understand exactly how the Fast Fourier Transform works, let's first remind ourselves of how the Discrete Fourier Transform works. The DFT, X, of the frequency index, k, for the signal, x, is given by Equation 26.

$$X_k = \sum_{n=0}^{N-1} x_n \cdot \left[\cos\left(2\pi \frac{k}{N} n\right) - i \cdot \sin\left(2\pi \frac{k}{N} n\right) \right]$$

Equation 26

We can understand this equation as a list of instructions:

1. Multiply each sample in the signal, x_n, by $\cos(2\pi \frac{k}{N} n)$ at frequency index $k=0$, producing a multiplied cosine wave.
2. Add all the points on the multiplied cosine wave together, producing the cosine component for frequency index $k=0$.
3. Multiply each sample in the signal, x_n, by $-\sin(2\pi \frac{k}{N} n)$ at frequency index $k=0$, producing a multiplied sine wave.
4. Add all the points on the multiplied sine wave together, producing the sine component for frequency index $k=0$.
5. Repeat steps 1–4, incrementing k each time until k = N−1.

The signal in Figure 102 contains eight samples, x_0–x_7. Therefore, to find the DFT for this signal, we have to multiply it by cosine waves and sine waves at eight different frequencies.

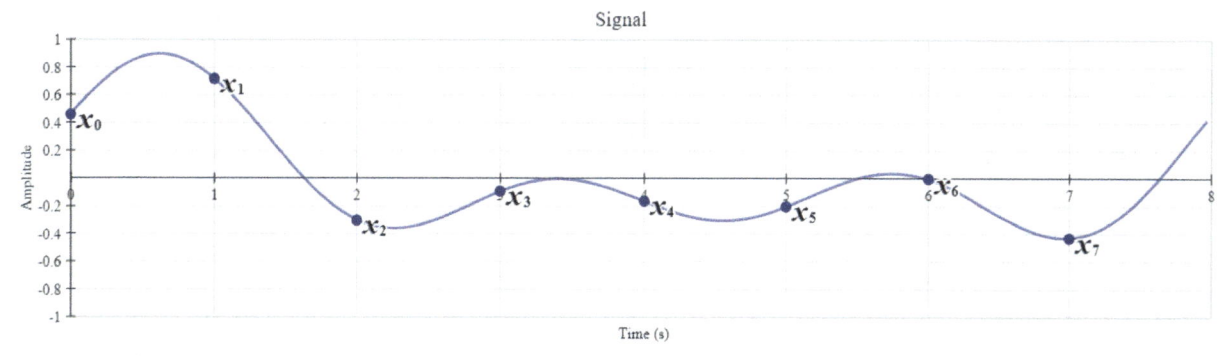

Figure 102

How the Fourier Transform works

Chapter 5

When we sample a cosine or sine wave using a number of samples that is a power of 2, something interesting happens as the frequency of the wave increases. Certain combinations of samples maintain consistent amplitudes, even though the frequency of the wave changes. As a result, when we multiply the signal by the cosine or sine wave at these specific sample points, the outcome remains constant even though the frequency of the cosine or sine wave has increased.

Consider the cosine waves in Figure 103 to Figure 106. In each figure, I have marked 2 samples, c_0 and c_4. See how even though the frequency of the cosine wave increases by 2 cycles from one graph to the next, the amplitudes of these samples remain the same.

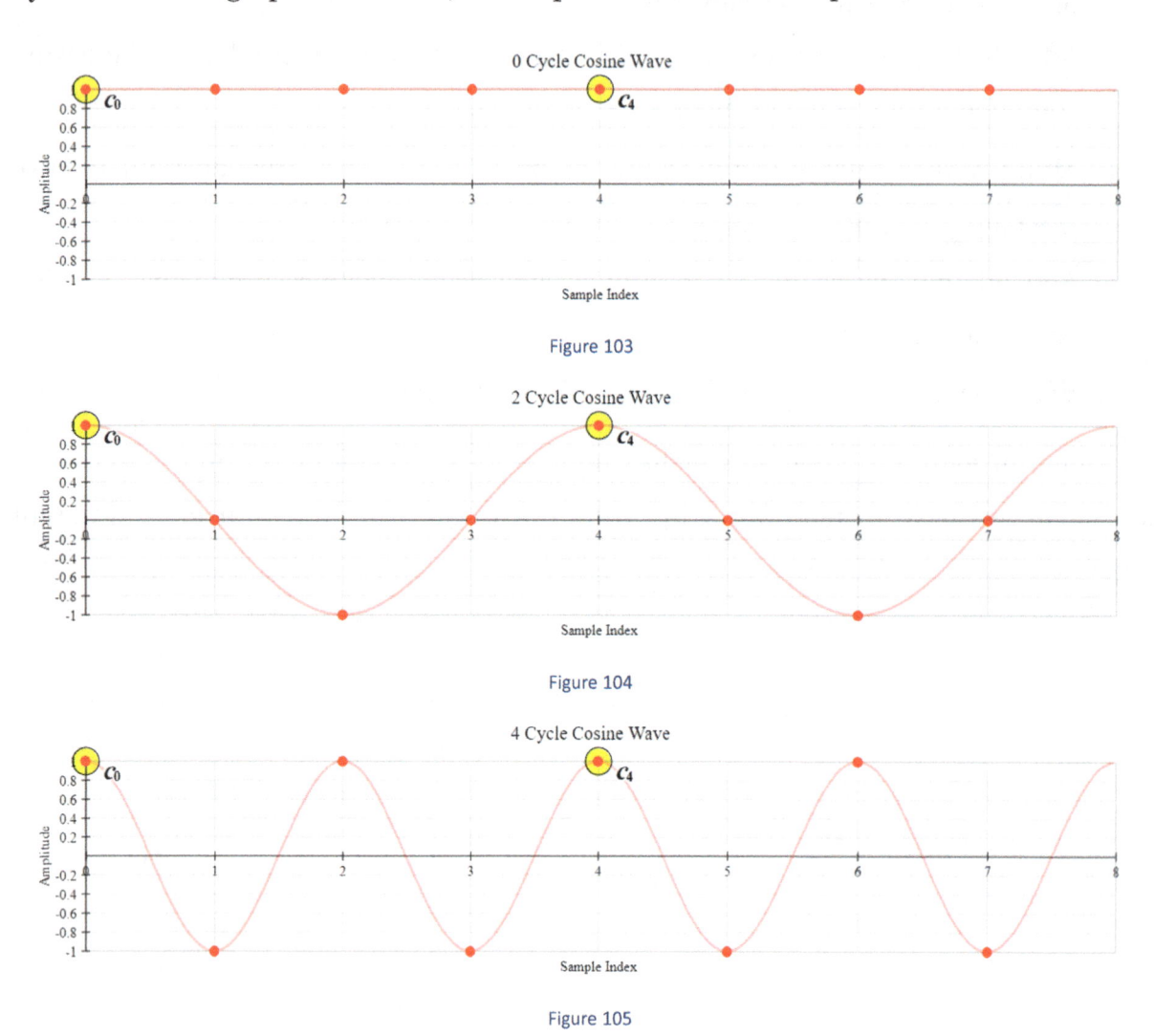

Figure 103

Figure 104

Figure 105

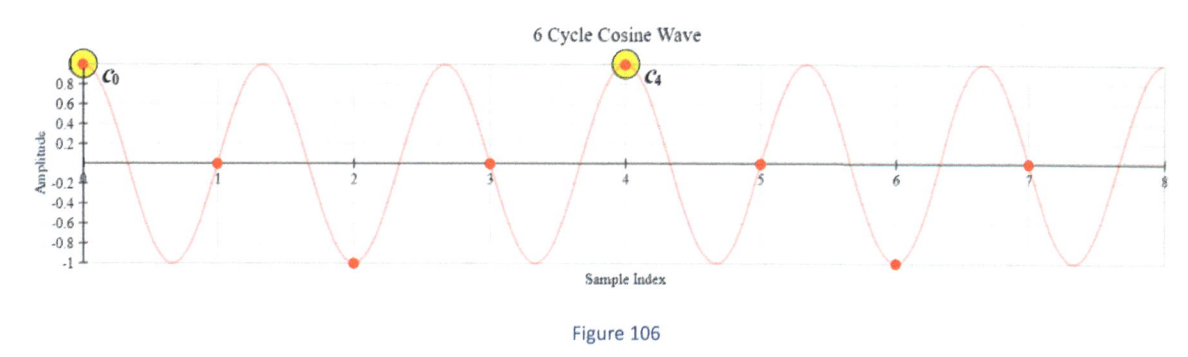

Figure 106

Given that this is the case, when we multiply x_0 by c_0, or x_4 by c_4, it doesn't matter which of the four frequencies we test; the answer will always be the same.

The same is true for the sine waves shown in Figure 107 to Figure 110. Even though the frequency of the sine wave increases by 2 cycles from one graph to the next, the amplitudes of samples s_0 and s_4 remain the same.

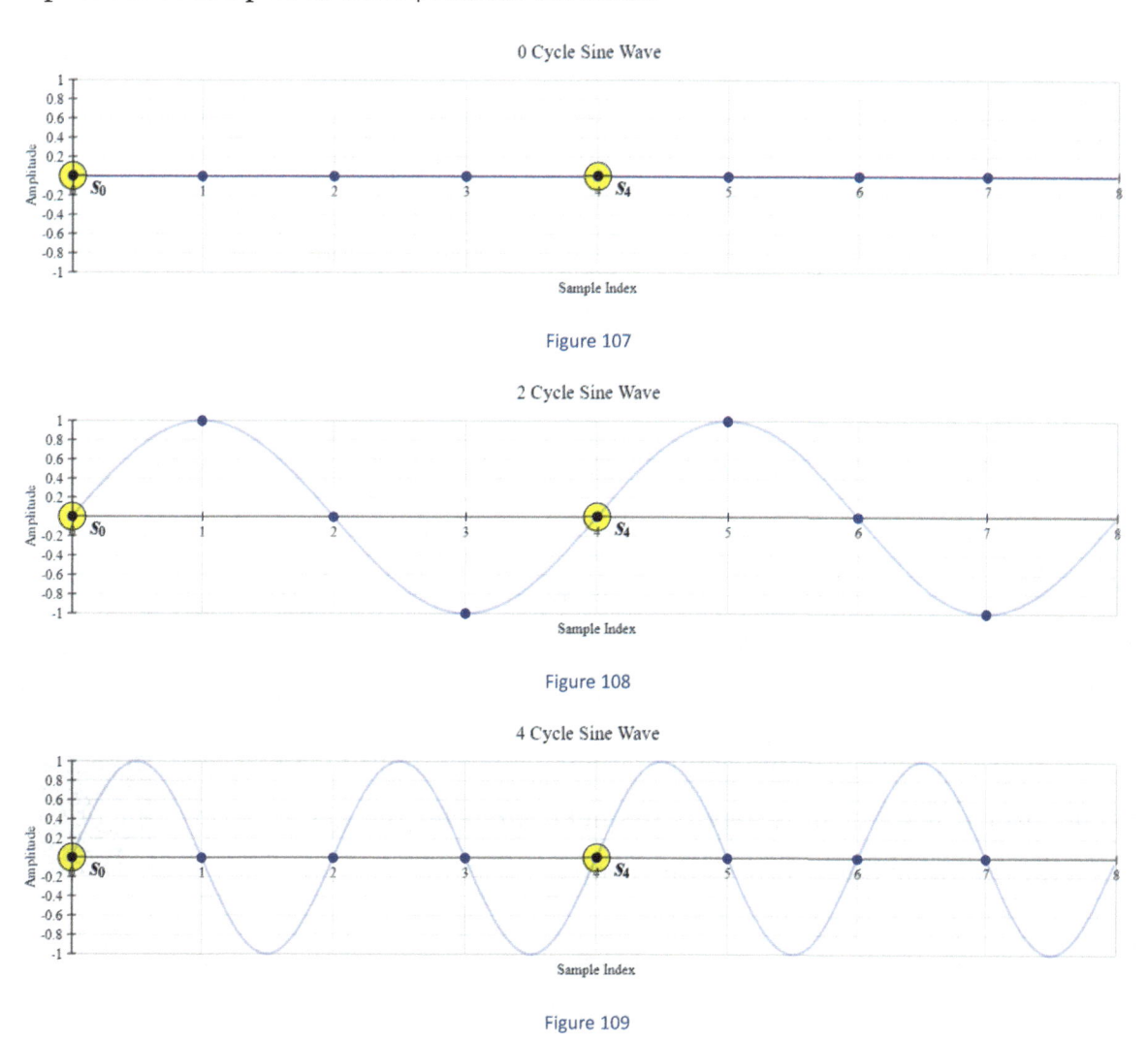

Figure 107

Figure 108

Figure 109

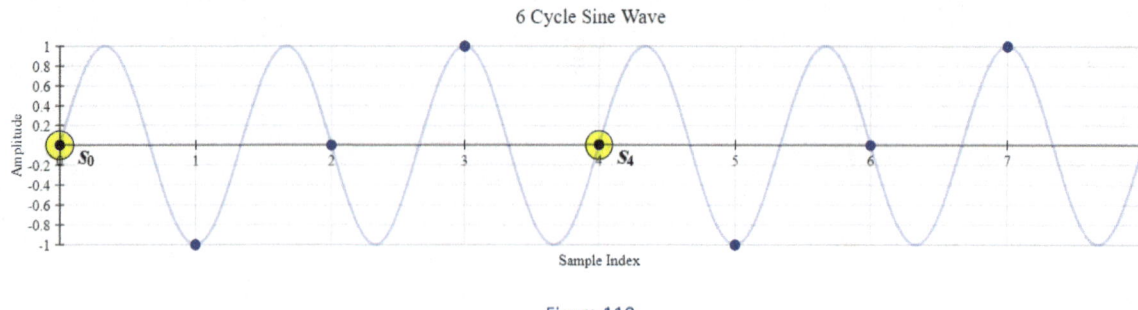

[Figure: 6 Cycle Sine Wave plot with samples s_0 and s_4 highlighted]

Figure 110

This phenomenon of amplitudes remaining constant over a number of different frequencies is true for other sample combinations too, the only difference being the number of frequencies for which the same amplitudes repeat. For example, in Figure 111 and Figure 112, c_2 and c_6 both have an amplitude of -1. This particular case repeats only twice out of the eight cosine waves tested.

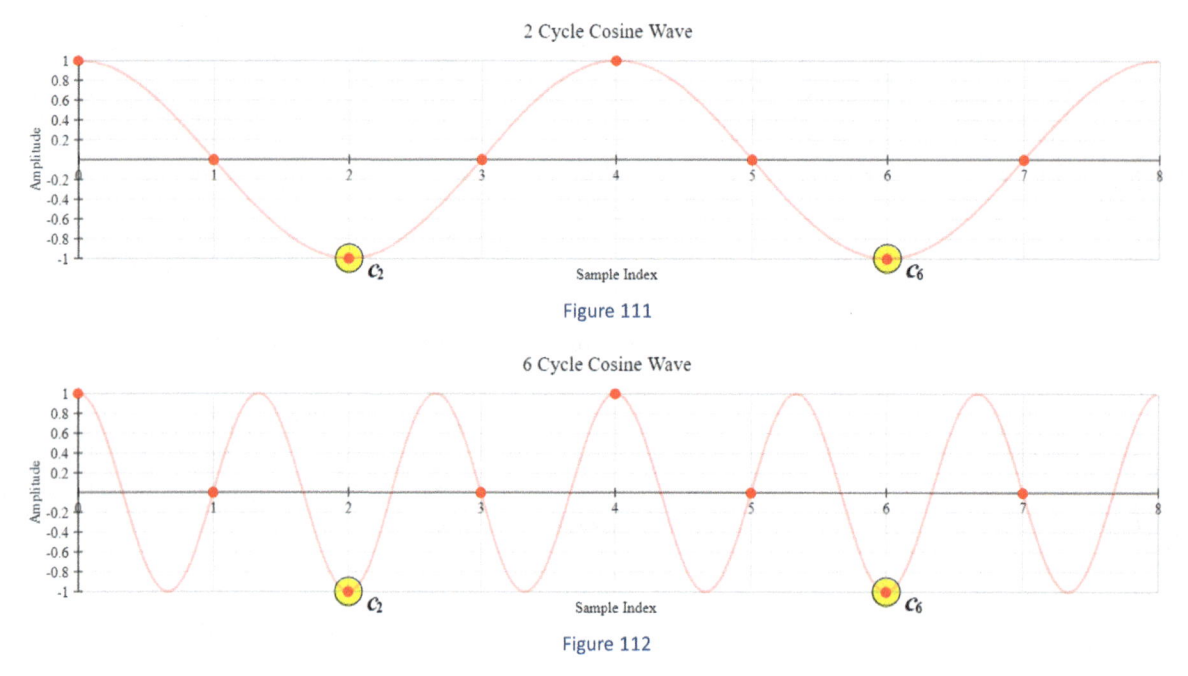

[Figure: 2 Cycle Cosine Wave with c_2 and c_6 highlighted]

Figure 111

[Figure: 6 Cycle Cosine Wave with c_2 and c_6 highlighted]

Figure 112

Can you find any more sample combinations whose amplitudes remain constant over a number of different frequencies? Scan the QR code or click on the following link to search for them.

Divide-and-conquer

A divide-and-conquer algorithm recursively breaks down a problem into two or more sub-problems of the same or related type until these become simple enough to be solved directly. The solutions to the sub-problems are then combined to give a solution to the original problem.

- Wikipedia

The FFT repeatedly divides an array of samples into smaller and smaller groups, ordering those samples in such a way that it can take advantage of the repeating nature of calculations in the DFT. It then conquers the problem by performing a simple calculation on each group, before combining the results recursively until it has calculated the DFT for the entire array of samples. We'll see how this works a little later.

Sorting an array of numbers

Before we consider the FFT, we'll see how divide-and-conquer works by considering a simpler problem. Then we'll take what we've learned and apply it to the FFT. The task is to sort an array of eight random numbers in ascending order.

28	6	73	96	63	71	59	62

Figure 113

Many algorithms have been proposed over the years, but the most efficient employ the divide-and-conquer approach. The sorting algorithm we'll be considering here is called Merge Sort.

We repeatedly divide the array in half until it cannot be divided any more. In order to keep the analogy consistent with the way the FFT uses the divide-and-conquer method, I've purposely made the number of elements in Figure 113 a power of 2. However, the merge-sort algorithm can work on arrays of any length.

Stage 1 – Divide

| 28 | 6 | 73 | 96 | 63 | 71 | 59 | 62 |

Figure 114 – The unsorted array

We divide the array of items to sort in two, producing Figure 115

| 28 | 6 | 73 | 96 | 63 | 71 | 59 | 62 |

Figure 115 - Divide into 2 groups of 4

... then divide it again, producing Figure 116

| 28 | 6 | 73 | 96 | 63 | 71 | 59 | 62 |

Figure 116 - Divide into 4 groups of 2

... and again, until we cannot divide it any more, producing Figure 117

| 28 | 6 | 73 | 96 | 63 | 71 | 59 | 62 |

Figure 117 - Divide into 8 groups of 1

Stage 2 – Conquer

Now we're going to conquer the problem. There are 8 groups in Figure 117. Each group contains a single element.

Starting with the first two groups, we label them Group A and Group B, as shown in Figure 118.

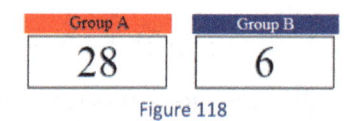

We compare the first element in Group A with the first element in Group B, as shown in Figure 119.

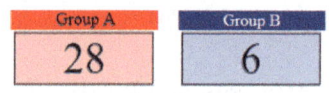

It may seem strange to say: "the first element" at this stage. There is only a single element in each group. However, as the process progresses, there will be more and more elements in each group, as we will see later.

The element in Group B is smaller, so we remove it from Group B and append it to a new group called Group C, as shown in Figure 120.

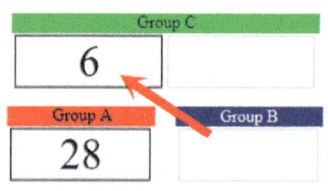

Group B is now empty, so we can append the remaining values in Group A to Group C, as shown in Figure 121.

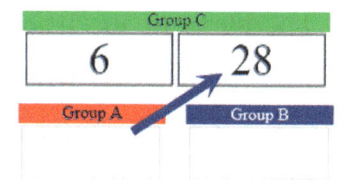

We repeat this process for each pair of groups in Figure 117.

We're now left with 4 groups, each containing two sorted elements, as shown in Figure 122.

| 6 | 28 | 73 | 96 | 63 | 71 | 59 | 62 |

Figure 122 – Four groups containing 2 sorted elements in each group

Again and again, we label the groups A and B, repeatedly comparing the first items in each group, and moving the smaller item into Group C.

We label the first group in Figure 122 Group A and the second, Group B.

We compare the first element in Group A with the first element in Group B, as shown in Figure 124.

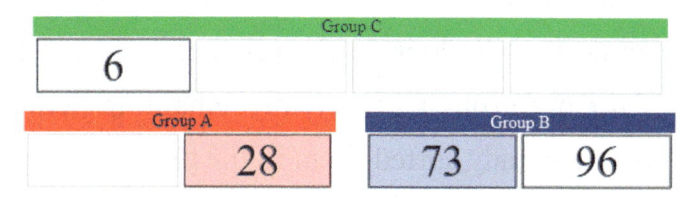

The element in Group A is smaller, so we remove it from Group A and append it to a new group called Group C, as shown in Figure 125.

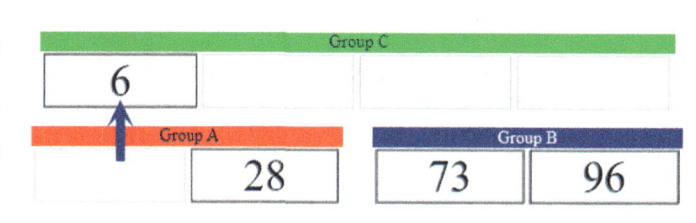

We have already removed the 6 from Group A, therefore the 28 is now the first element in that group.

We compare the first element in Group A with the first element in Group B, as shown in Figure 126.

The element in Group A is smaller, so we remove it from Group A and appended to Group C, as shown in Figure 127.

Figure 127

As Group A is now empty, we can move the remaining elements in Group B directly into Group C. There is no need to sort them, as we already sorted them in the previous stage.

Figure 128

We repeat this process for the second two groups in Figure 122, leaving us with 2 groups, each containing 4 sorted elements, as shown in Figure 129.

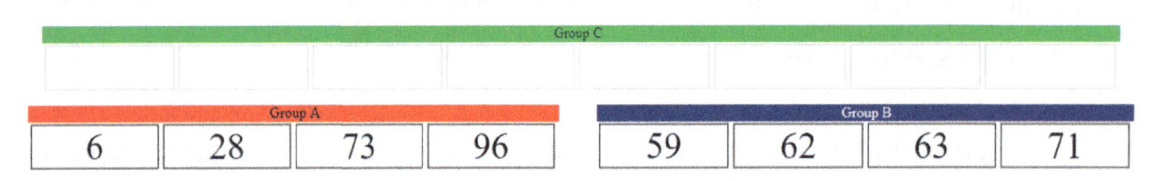

Figure 129 – Two groups containing 4 sorted elements in each group

We repeat the process of comparing the first elements in each group and moving the smaller element into Group C until all the elements have been sorted, producing the final sorted array shown in Figure 130.

Figure 130

To see the entire process animated from start to finish, scan the QR code opposite or click on the following link: Merge-sort animation.

The advantage of the divide-and-conquer method is that each new stage builds on the work already done in the previous stage, without having to do the same work again. Once one group is empty, any remaining elements in the other group need not be sorted, as we already sorted them in the previous stage. In the worst case, for very large arrays, the total number of operations required by this method approaches $N \log_2(N)$, where N is the number of elements to be sorted.

For example, an array containing 256 elements would only require a maximum of 2048 operations to sort the array, as opposed to the 65,536 operations using a sorting algorithm which didn't employ the divide-and-conquer method.

The same principle of splitting the samples into groups of similar calculations, and doing some of the work in one stage so that we need not repeat the same work in the next, is what makes the FFT so computationally efficient. Over the next few sections, we'll be coming back to the principles we've just learned from the merge-sort algorithm as we apply the divide-and-conquer method to the Discrete Fourier Transform.

Divide-and-conquer in the FFT

In the section called Repeating calculations, we saw how the results from the calculation of certain groups of samples in the DFT repeat themselves if we ensure that the number of samples is a power of 2. We're now going to go back to the sampled signal and see how we can group those samples together to take advantage of the repeating nature of the calculations.

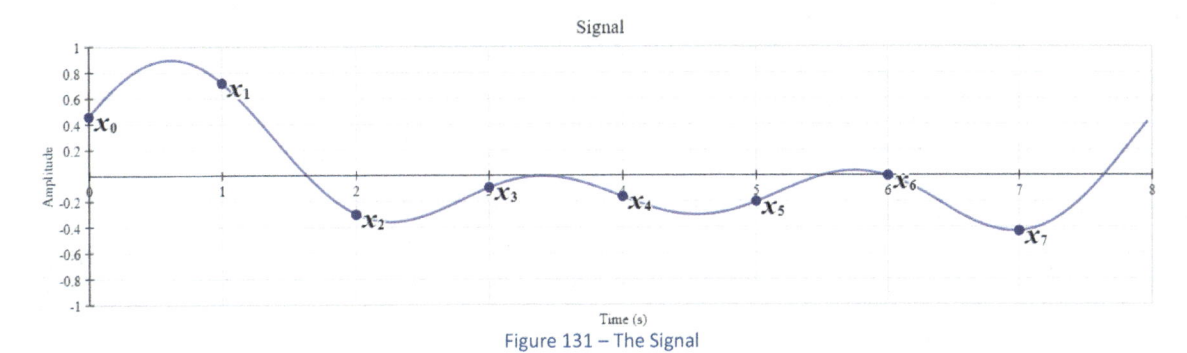

Figure 131 – The Signal

Divide stage

The FFT algorithm repeatedly divides the elements of the input array into smaller and smaller arrays, just like the merge-sort algorithm. However, unlike merge sort, it orders them a little differently. Instead of grouping consecutive elements, like merge sort does, the FFT groups the even and odd elements together in each stage. Figure 131 shows the signal on which we're going to perform the FFT algorithm. It contains 8 samples, labeled x_0 to x_7.

Figure 132 shows the amplitudes of the samples x_0 to x_7.

x_0	x_1	x_2	x_3	x_4	x_5	x_6	x_7
0.46	0.72	-0.3	-0.09	-0.16	-0.2	0	-0.43

Figure 132 – Sample values for the signal in Figure 131

I have colored all the even-numbered samples, x_0, x_2, x_4, and x_6 in red, and all the odd-numbered samples, x_1, x_3, x_5, and x_7 in blue. This is because the first operation of the divide stage is to split these samples into two groups. The first group contains only the even-numbered samples, and the second group contains only the odd-numbered samples, as shown in Figure 133.

x_0	x_2	x_4	x_6	x_1	x_3	x_5	x_7
0.46	-0.3	-0.16	0	0.72	-0.09	-0.2	-0.43

Figure 133 – The sample array divided into even and odd groups

The two groups are divided again so that now there are 4 groups containing 2 samples per group, as shown in Figure 134.

x_0	x_4	x_2	x_6	x_1	x_5	x_3	x_7
0.46	-0.16	-0.3	0	0.72	-0.2	-0.09	-0.43

Figure 134 – The sample array divided again

The divide stage is complete once the array has been divided enough times so that there are only 2 samples in each group. Unlike the merge-sort algorithm, in the FFT there is no need to divide any further, as the DFT works on pairs of samples.

To see this process animated, scan the QR code opposite, or click on the following link: FFT Divide Animation.

Conquer stage

Let's go back to the DFT equation and see how dividing the array of samples up into groups containing only 2 samples per group makes calculating the DFT for each group very easy.

$$X_k = \sum_{n=0}^{N-1} x_n \cdot \left[cos\left(2\pi \frac{k}{N} n\right) - i \cdot sin\left(2\pi \frac{k}{N} n\right) \right]$$

Equation 27 - The DFT Equation

To see what the DFT is doing, we're going to run it on the signal in Figure 131. In Equation 27, n is the sample index and x_n represents the signal. You can see the amplitudes for each sample x_n in Figure 134. Each sample must be multiplied by its corresponding sample on a cosine wave and on an inverted sine wave. The sine wave is inverted on account of the minus sign between the cosine term and the sine term in the DFT as shown in Equation 27.

Calculating a 2-point DFT

In the collection of graphs in Figure 135, I've drawn the signal in green, the cosine wave in blue, and the sine wave in red.

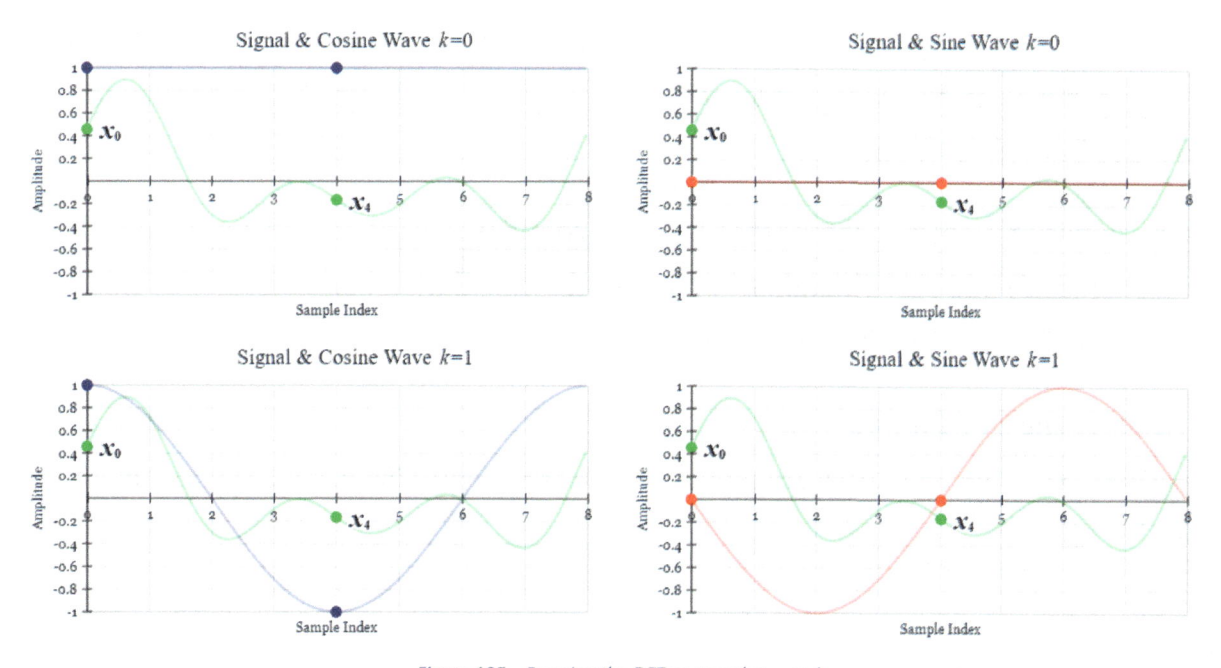

Figure 135 – Running the DFT on samples x_0 and x_4

There are only 2 samples in each group, so $N=2$. The sample index, n, ranges from 0 to 1. In the FFT, this sample index can be a little confusing. Remember that the divide stage of the FFT reorders the samples so that we end up with the groups of even and odd samples shown in Figure 134. The first group contains the samples x_0 and x_4 (not x_0 and x_1). These are the green dots on the signal which are labeled on each of the graphs in Figure 135.

Even though they might be the first and fourth samples in the overall signal, as far as this 2-point DFT is concerned, x_0 and x_4 are the only samples that exist, so for sample x_0, $n=0$, and for sample x_4, $n=1$.

In the DFT, the number of frequencies it will test is equal to the number of samples. Therefore, the 2 samples in each group will be tested at two frequencies, $k=0$ (the graphs on the top row in Figure 135) and $k=1$ (the graphs on the bottom row in Figure 135).

At each frequency, the DFT multiplies the signal by a cosine wave (the graphs on the left) and an inverted sine wave (the graphs on the right). It then sums the results of all the multiplications to give the DFT of the signal for each of the frequencies. So the DFT, X, of the signal at frequency index $k=0$ will be $X_{k=0}$, and the DFT of the signal at frequency index $k=1$ will be $X_{k=1}$.

The imaginary number, i, shows that this is a complex calculation, so we need to keep the cosine (real) and sine (imaginary) parts of the calculation separate. Let us calculate them for the first 2-point DFT of our signal.

Calculating the DFT for $k=0$

The DFT, $X_{k=0}$, for the first frequency index, $k=0$, can be calculated as shown in Equation 28:

$$X_{k=0} = x_0 \cdot \left[cos\left(2\pi \cdot \frac{0}{2} \cdot 0\right) - i \cdot sin\left(2\pi \cdot \frac{0}{2} \cdot 0\right) \right] + x_4 \cdot \left[cos\left(2\pi \cdot \frac{0}{2} \cdot 1\right) - i \cdot sin\left(2\pi \cdot \frac{0}{2} \cdot 1\right) \right]$$

Equation 28 – Calculating the DFT for X_0

For the first frequency index, $k=0$. This is the red number in the equation above. The index, n, ranges from 0 to 1 and is colored blue.

We can simplify Equation 28 to give Equation 29.

$$X_{k=0} = x_0 \cdot [cos(0) - i \cdot sin(0)] + x_4 \cdot [cos(0) - i \cdot sin(0)]$$

Equation 29

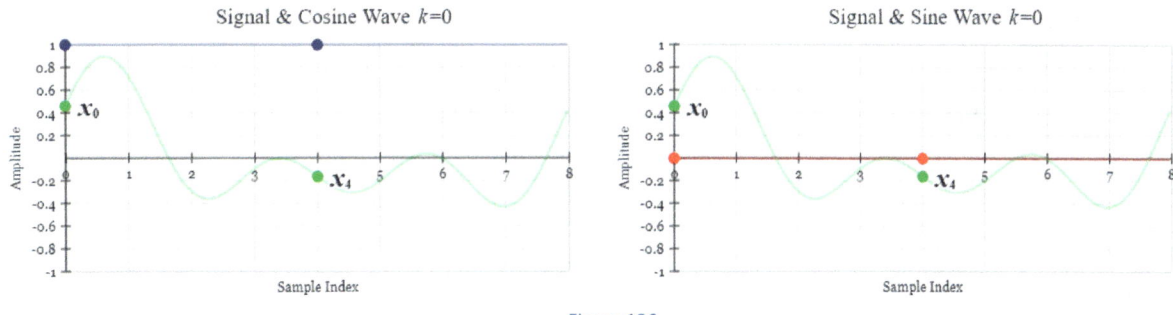

Figure 136

We can see from Figure 136 that it doesn't matter what the sample index, *n*, is. At *k*=0, the amplitude of the cosine wave, shown by the blue dots, is always 1, and the amplitude of the sine wave, shown by the red dots, is always 0. We can understand this mathematically from Equation 29, because:

$$cos\,(0) = 1 \qquad -sin\,(0) = 0$$

So we can further simplify Equation 29, giving us Equation 30.

$$X_{k=0} = x_0 + x_4$$

Equation 30

Reading off the amplitudes of samples x_0 and x_4 from Figure 134, we can use this equation to calculate $X_{k=0}$ for our signal, giving us Equation 31.

$$X_{k=0} = 0.46 + -0.16 = 0.3$$

Equation 31

Calculating the DFT for *k*=1

The DFT, $X_{k=1}$, for the second frequency index, *k*=1, can be calculated as shown in Equation 32.

$$X_{k=1} = x_0 \cdot \left[cos\left(2\pi \cdot \frac{1}{2} \cdot 0\right) - i \cdot sin\left(2\pi \cdot \frac{1}{2} \cdot 0\right) \right] + x_4 \cdot \left[cos\left(2\pi \cdot \frac{1}{2} \cdot 1\right) - i \cdot sin\left(2\pi \cdot \frac{1}{2} \cdot 1\right) \right]$$

Equation 32

For the second frequency index, *k*=1. This is the red number in the equation above. The index, *n*, ranges from 0 to 1, and is colored blue.

How the Fourier Transform works

We can simplify Equation 32 to give Equation 33.

$$X_{k=1} = x_0 \cdot [cos(0) - i \cdot sin(0)] + x_4 \cdot [cos(\pi) - i \cdot sin(\pi)]$$

Equation 33

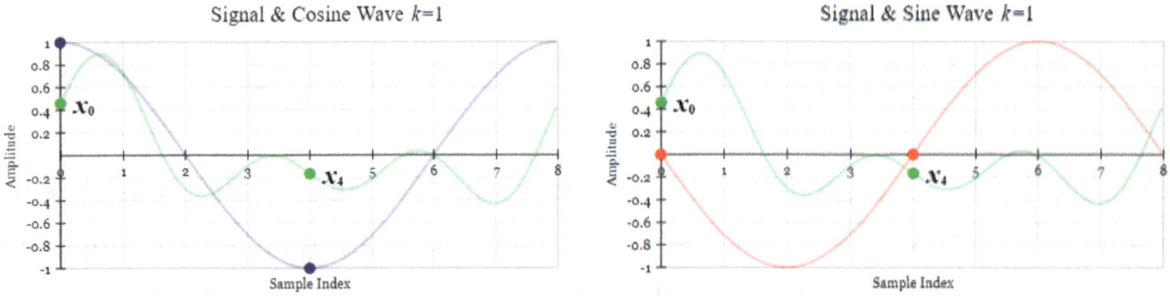

Figure 137

From the left-hand graph in Figure 137, we can see that, for x_0, the amplitude of the corresponding sample on the cosine wave is 1. From the right-hand graph, we can see that the amplitude of the corresponding sample on the sine wave is 0. For x_4, the amplitude of the corresponding sample on the cosine wave is -1, and the amplitude of the corresponding sample on the sine wave is 0. We can understand this mathematically from Equation 33, because:

$cos(0)=1$ $-sin(0)=0$ $cos(\pi)=-1$ $-sin(\pi)=0$

So we can further simplify Equation 33, giving us Equation 34.

$$X_{k=1} = x_0 - x_4$$

Equation 34

Reading off the amplitudes of samples x_0 and x_4 from Figure 134, we can use this equation to calculate $X_{k=1}$ for our signal, giving us Equation 35.

$$X_{k=1} = 0.46 - -0.16 = 0.62$$

Equation 35

Repeating the process for the other sample groups

Conquering a 2-point DFT (a DFT with only 2 samples in it) is very easy. All we need to do is to add the samples for the first frequency term and subtract them for the second. We can now apply this method to each group of samples from the end of the divide stage. This process can be represented graphically using a Butterfly Diagram.

The butterfly diagram

At the heart of the FFT algorithm sits a butterfly. We're not talking about a real butterfly, of course, but a mathematical one. The shape of the data-flow diagram for a 2-point DFT is reminiscent of a butterfly's wings.

The FFT is a recursive algorithm. This means that the core process of conquering groups of samples that we saw in the previous section will be repeated many times on larger and larger sample groups.

A butterfly diagram, like the one in Figure 138, is often referred to as the inner butterfly. It provides a simple way of visualizing this core process. As the algorithm advances, it will repeat the single butterfly many times, interleaving it with other butterflies as we combine more and more samples in each group.

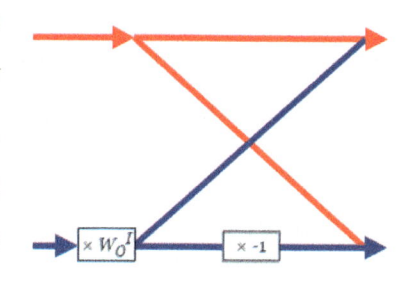

Figure 138 – A Single Butterfly

The diagram comprises the following symbols:

→ (orange)	Even sample input
→ (blue)	Odd sample input
→ (orange/blue)	Addition of two samples
→ (blue/orange)	Addition of two samples
× -1	Multiplication by -1
× W_0^1	Multiplication by a twiddle factor

The 2-point butterfly

Using these symbols, we can build butterfly diagrams to conquer each of the four groups of samples shown in Figure 139, and calculate a 2-point DFT for each group.

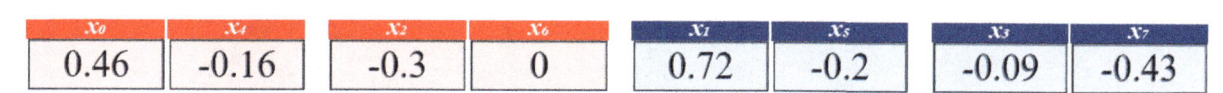

x_0	x_4	x_2	x_6	x_1	x_5	x_3	x_7
0.46	-0.16	-0.3	0	0.72	-0.2	-0.09	-0.43

Figure 139

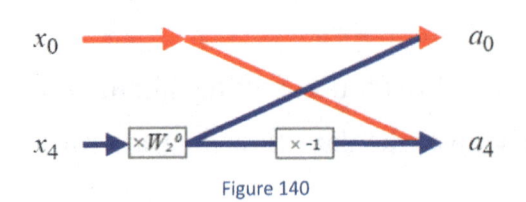

Figure 140

Figure 140 shows the butterfly diagram which conquers the first group of samples from Figure 139, x_0 and x_4. The sample x_0 enters the butterfly at the top left.

The sample x_4 enters the butterfly at the bottom left, where it is immediately multiplied by the twiddle factor, W_2^0. We can disregard the twiddle factor at this stage as it is equal to 1. We'll find out what a twiddle factor is in the next section.

As we're about to perform many 2-point DFTs on many different samples, we need to change our notation. Rather than calling the result of the DFT $X_{k=l}$, where l is the frequency index, the results of all the 2-point DFTs we're about to calculate will be called a_n, where n is the index of the sample that has flowed horizontally through the butterfly. We can see this in Figure 140.

We add x_0 to x_4 at the top right of the butterfly to produce a_0, as shown in Equation 36.

$$a_0 = x_0 + x_4$$

Equation 36

We multiply x_4 by the twiddle factor, W_2^0, which is equal to 1. Then it is multiplied by -1 and added to x_0 at the bottom right of the butterfly to produce a_4, as shown in Equation 37.

$$a_4 = x_0 + x_4 \times W_2^0 \times -1$$

Equation 37

We can simplify Equation 37, giving Equation 38.

$$a_4 = x_0 - x_4$$

Equation 38

We repeat the same idea for all the groups using the butterflies shown in Figure 141 to Figure 144, producing Equation 39 to Equation 46.

How the Fourier Transform works — Chapter 5

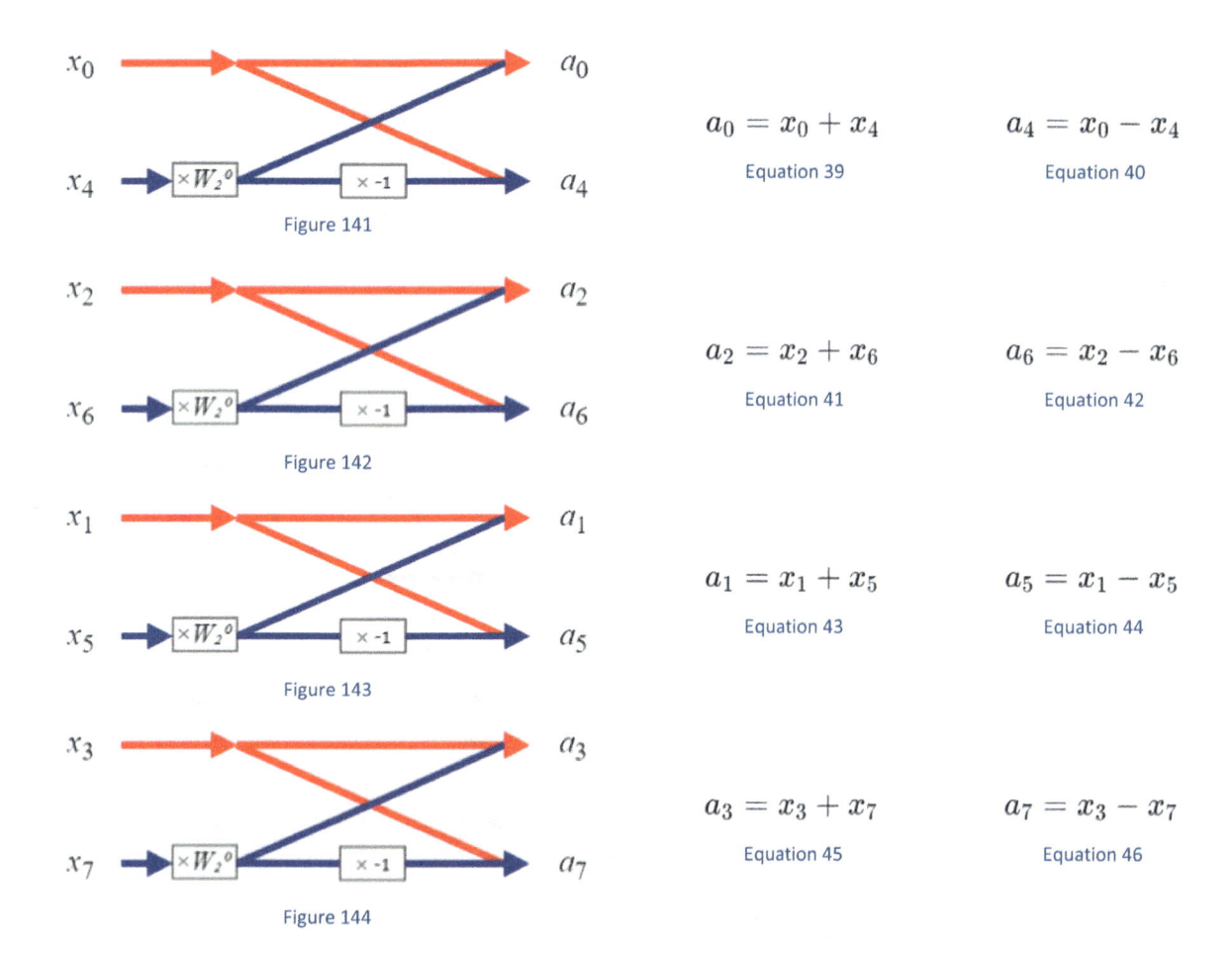

Figure 141

$$a_0 = x_0 + x_4$$
Equation 39

$$a_4 = x_0 - x_4$$
Equation 40

Figure 142

$$a_2 = x_2 + x_6$$
Equation 41

$$a_6 = x_2 - x_6$$
Equation 42

Figure 143

$$a_1 = x_1 + x_5$$
Equation 43

$$a_5 = x_1 - x_5$$
Equation 44

Figure 144

$$a_3 = x_3 + x_7$$
Equation 45

$$a_7 = x_3 - x_7$$
Equation 46

The 4-point butterfly

Now we're going to combine these four groups of 2-point butterflies into two groups of 4-point butterflies to calculate two 4-point DFTs for this signal, as shown in Figure 145. This time, the result of the DFT is denoted by b_n. Again, n is the index of the sample that has flowed horizontally through the butterfly.

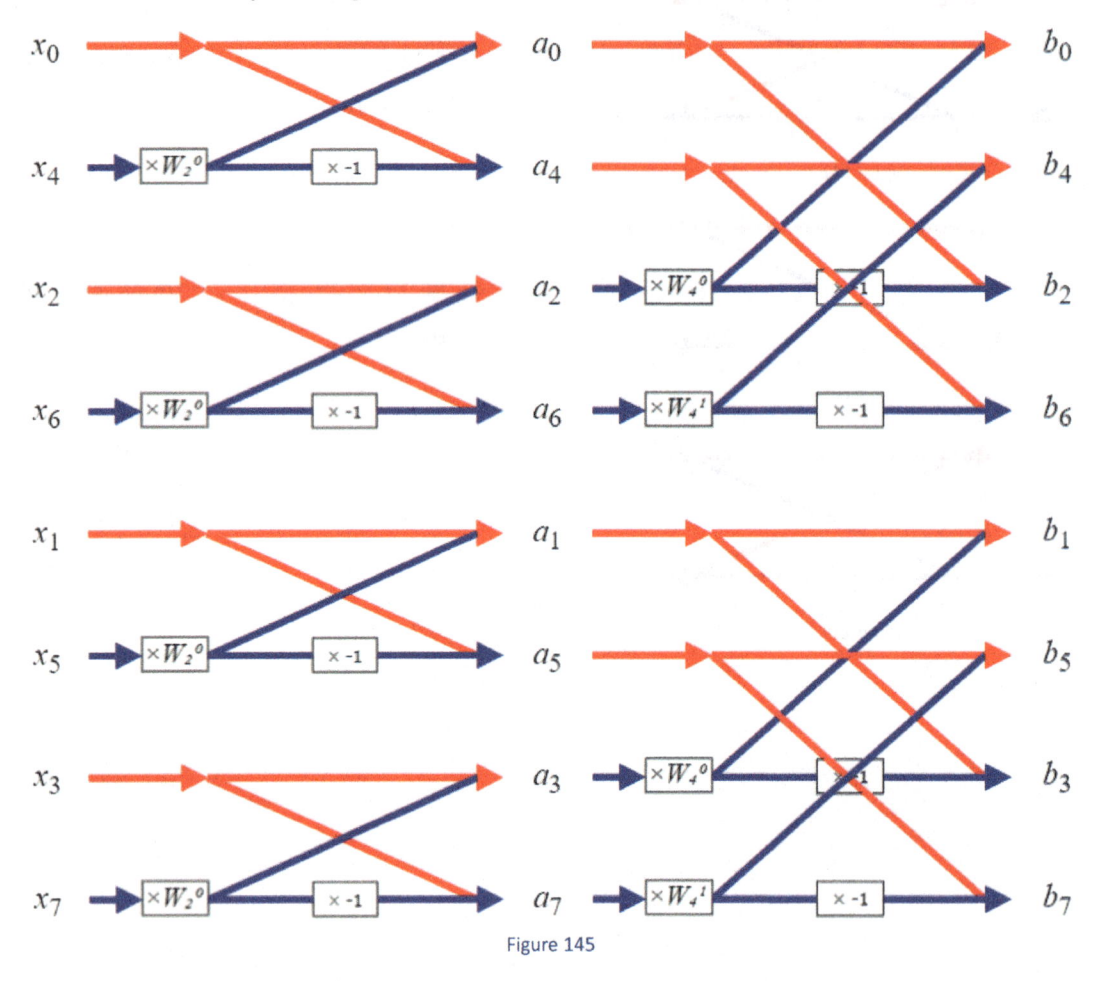

Figure 145

The FFT uses the outputs of the 2-point butterflies as inputs to the 4-point butterflies as shown in Figure 145. This way, when calculating the 4-point DFTs, it saves having to redo calculations it already performed, making the calculation a lot more efficient.

To demonstrate this, let's look at how we would calculate a 4-point DFT the long way round using individual samples as inputs, compared with how the FFT calculates it by using the outputs of two 2-point DFTs as inputs.

Method 1: The long way round

A 4-point DFT splits the signal into 4 cosine components and 4 sine components at 4 different frequencies. The signal, cosine, and sine waves for the first group of 4-point DFTs, involving samples x_0, x_2, x_4, and x_6, are shown in Figure 146.

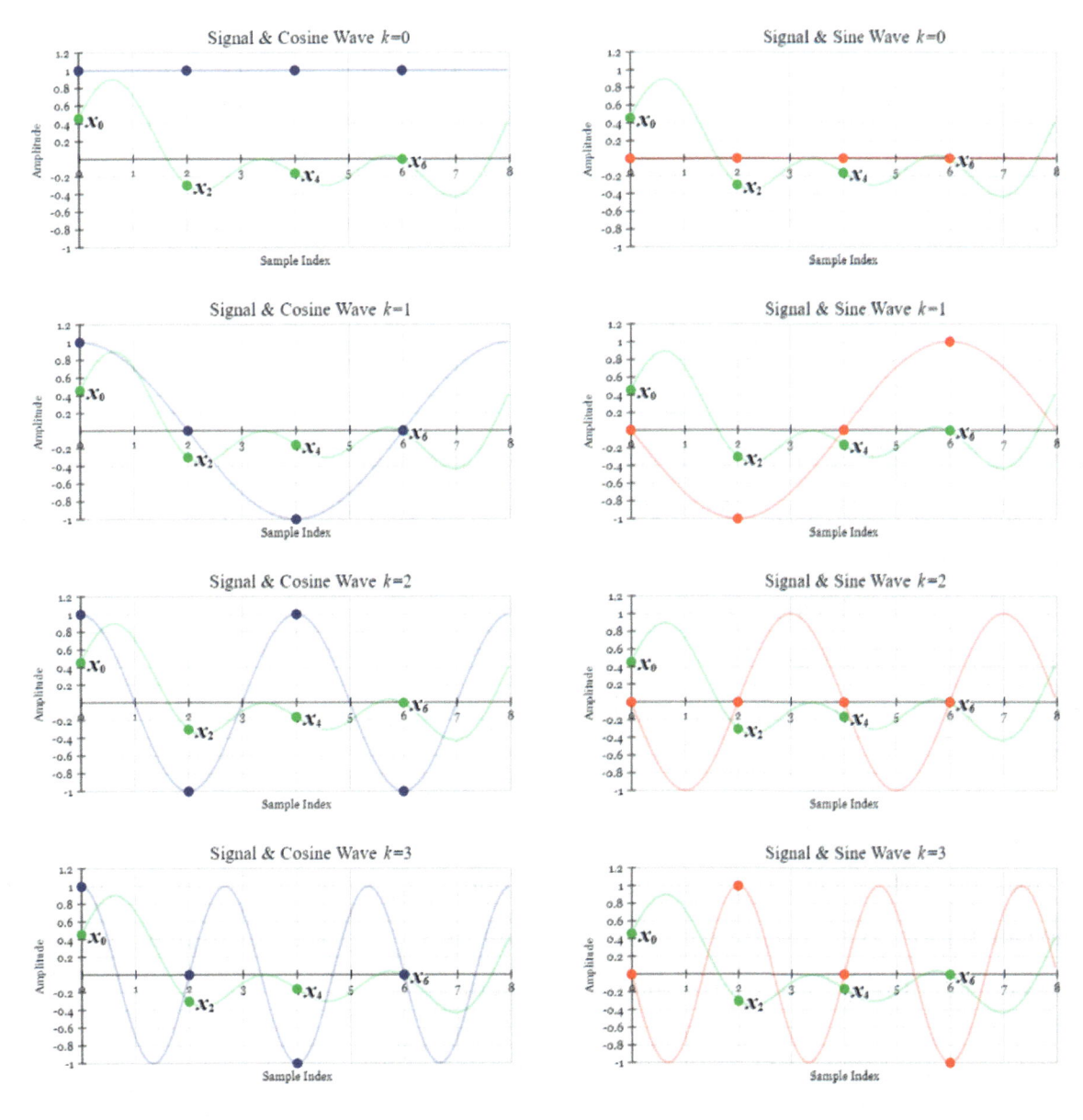

Figure 146

There are four rows of graphs. Each row contains a cosine wave (left-hand graph) and an inverted sine wave (right-hand graph) at one frequency to be tested. Let's demonstrate how to calculate the DFT for one of these frequencies.

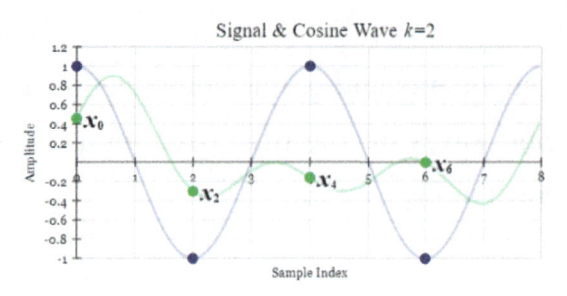

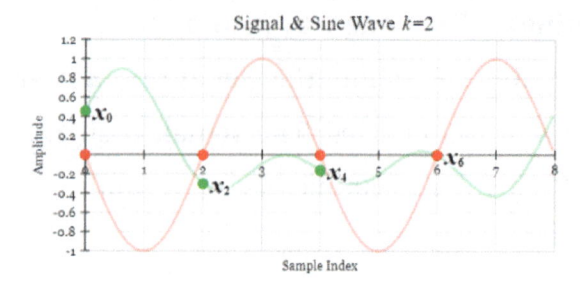

Figure 147

The graphs in Figure 147 show the signal, cosine, and sine waves for the frequency index *k=2*. At this frequency, the cosine and sine waves will oscillate twice over the period of the signal. Let's substitute that into the DFT equation, together with all the other information we have about the signal.

$$X_k = \sum_{n=0}^{N-1} x_n \cdot \left[\cos\left(2\pi \frac{k}{N} n\right) - i \cdot \sin\left(2\pi \frac{k}{N} n\right) \right]$$

Equation 47 - The DFT Equation

There are 8 samples in the signal. However, we have split the signal into two halves, each containing 4 samples, so *N=4*. The first group contains the 4 samples x_0, x_2, x_4, and x_6, so the index, *n*, ranges from 0 to 3. Remember that the imaginary number, *i*, shows that this is a complex calculation. The sine part is imaginary, so we have to keep the cosine and sine calculations separate. We'll split the calculation into two parts, one for the cosine component (the real part) and one for the sine component (the imaginary part).

Calculating the Cosine (Real) Part

We'll use the curly $\Re$ symbol in Equation 48 to denote that this is the real (cosine) part of the calculation. We'll use b_2 to denote the result of the DFT for this frequency to keep things consistent with the butterfly diagram in Figure 145.

$$\Re(b_2) = x_0 \times \cos\left(2\pi \cdot \frac{2}{4} \cdot 0\right) + x_2 \times \cos\left(2\pi \cdot \frac{2}{4} \cdot 1\right) + x_4 \times \cos\left(2\pi \cdot \frac{2}{4} \cdot 2\right) + x_6 \times \cos\left(2\pi \cdot \frac{2}{4} \cdot 3\right)$$

Equation 48

If we calculate the brackets for Equation 48, we get Equation 49.

$$\Re(b_2) = x_0 \times \cos(0) + x_2 \times \cos(\pi) + x_4 \times \cos(2\pi) + x_6 \times \cos(3\pi)$$

Equation 49

We see the 4 sample points in the cosine graph in Figure 148:

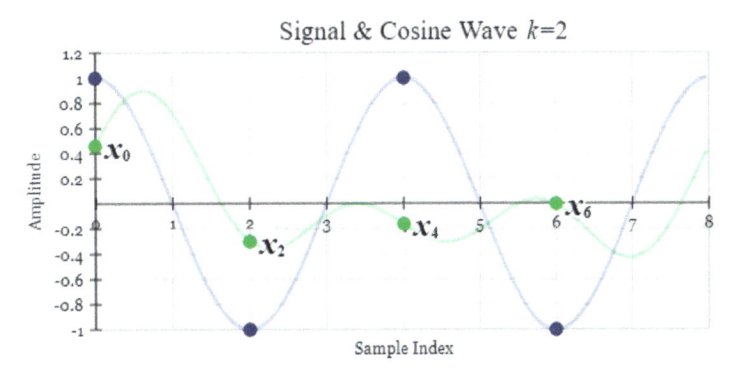

$$\cos(0)=1 \quad \cos(\pi)=-1 \quad \cos(2\pi)=1 \quad \cos(3\pi)=-1$$

Figure 148

The cosine component of the DFT for this frequency simplifies to Equation 50.

$$\Re(b_2) = (x_0 \times 1) + (x_2 \times -1) + (x_4 \times 1) + (x_6 \times -1)$$

Equation 50

Calculating the Sine (Imaginary) Part

We'll use the curly $\Im$ symbol in Equation 51 to denote that this is the imaginary part of the calculation. Again, b_2 denotes the result of the DFT at frequency index $k=2$.

$$\Im(b_2) = x_0 \times -sin\left(2\pi \cdot \frac{2}{4} \cdot 0\right) + x_2 \times -sin\left(2\pi \cdot \frac{2}{4} \cdot 1\right) + x_4 \times -sin\left(2\pi \cdot \frac{2}{4} \cdot 2\right) + x_6 \times -sin\left(2\pi \cdot \frac{2}{4} \cdot 3\right)$$

Equation 51

If we calculate the brackets for Equation 51, we get Equation 52.

$$\Im(b_2) = x_0 \times -sin(0) + x_2 \times -sin(\pi) + x_4 \times -sin(2\pi) + x_6 \times -sin(3\pi)$$

Equation 52

As we see from Figure 149, the values at the four sample points are:

- $-sin(0) = 0$
- $-sin(\pi) = 0$
- $-sin(2\pi) = 0$
- $-sin(3\pi) = 0$

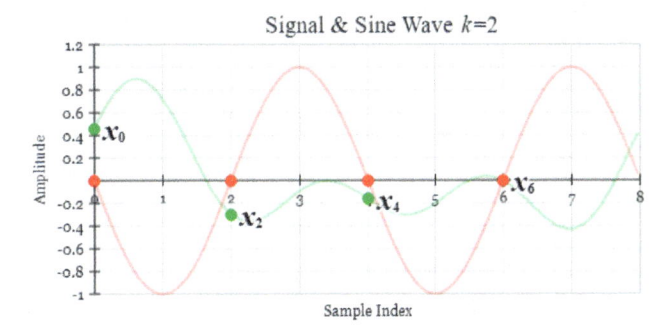

Figure 149

The sine component of the DFT at this frequency simplifies to Equation 53.

$$\Im(b_2) = (x_0 \times 0) + (x_2 \times 0) + (x_4 \times 0) + (x_6 \times 0)$$

Equation 53

All the samples in Equation 53 are multiplied by zero. Therefore, there is no sine (imaginary) component at this frequency.

Method 2: Building a 4-point DFT out of two 2-point DFTs

As the cosine (real) part of the DFT for this frequency is given by Equation 54…

$$\Re(b_2) = (x_0 \times 1) + (x_2 \times -1) + (x_4 \times 1) + (x_6 \times -1)$$

Equation 54

…and the sine (imaginary) part of the DFT for this frequency is zero, as we saw in Equation 53, we can say that the DFT for this frequency is entirely made up of the cosine component. Multiplying out the brackets for the cosine (real) part of the DFT for this frequency gives us Equation 55.

$$b_2 = x_0 - x_2 + x_4 - x_6$$

Equation 55

We can slightly rearrange Equation 55 to get Equation 56.

$$b_2 = (x_0 + x_4) - (x_2 + x_6)$$

Equation 56

Let us recall the 2-point DFTs we calculated in the previous section. I have put them together in a list in Figure 150.

$$a_0 = x_0 + x_4 \qquad a_4 = x_0 - x_4$$

$$a_2 = x_2 + x_6 \qquad a_6 = x_2 - x_6$$

$$a_1 = x_1 + x_5 \qquad a_5 = x_1 - x_5$$

$$a_3 = x_3 + x_7 \qquad a_7 = x_3 - x_7$$

Figure 150

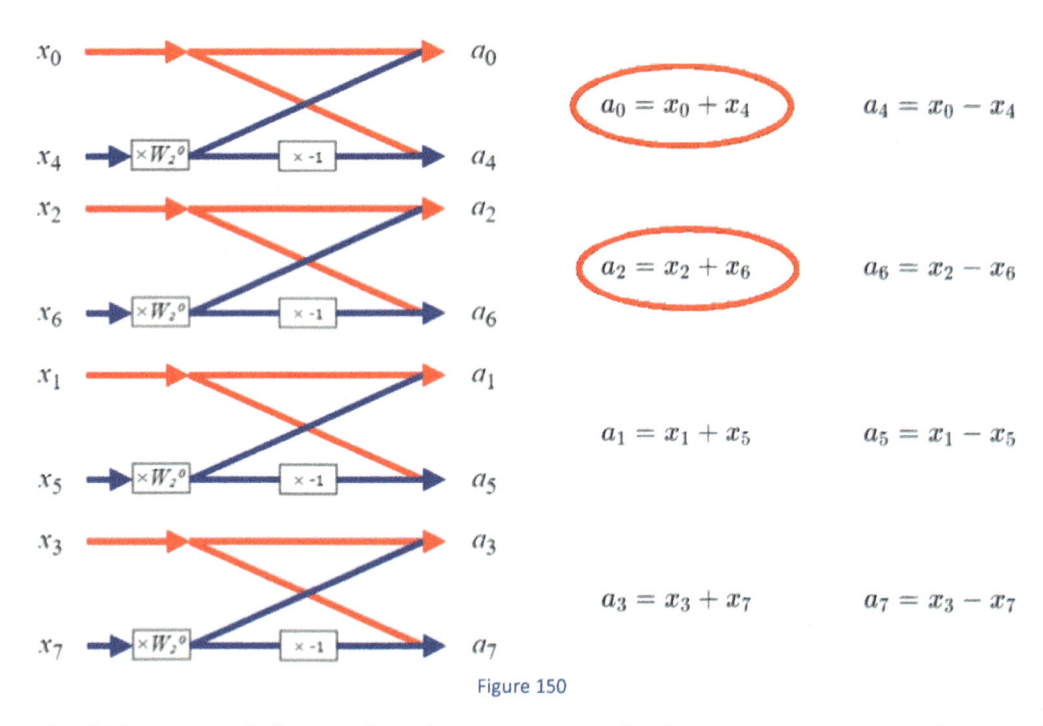

We already did some of the work. Thus we can calculate a 4-point DFT by combining two of the 2-point DFTs we calculated before, giving us Equation 57.

$$b_2 = a_0 - a_2$$

Equation 57

This is what makes the FFT so efficient. As we learned at the beginning of this chapter, by remembering the result it calculated at one frequency, the FFT can use that result when it occurs again at another frequency, without having to recalculate it.

This process can be represented graphically by interleaving two butterflies, as shown in Figure 151.

See how the results a_0 (first input to the interleaved butterflies) and a_2 (third input to the interleaved butterflies) from the 2-point DFTs are combined to give the DFT, b_2 (third output from the interleaved butterflies).

This all seems to work very well. So why do we need twiddle factors on inputs a_2 and a_6?

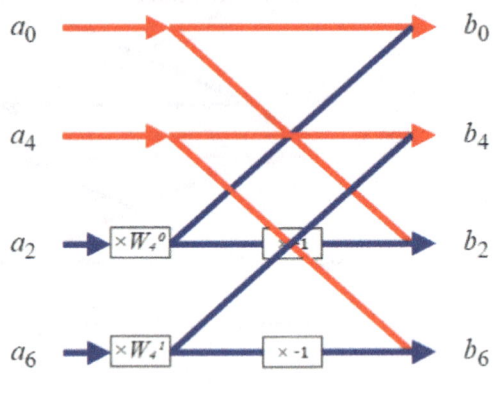

Figure 151

Twiddle factors

The beauty of the FFT algorithm is that it does the same thing over and over again. It treats every stage of the calculation in exactly the same way. However, this causes a problem. Not all the samples in the signal were sampled at the same time. They occupy different positions on the x-axis in relation to the cosine and sine waves. How can the FFT algorithm work if its "one-size-fits-all" approach doesn't correspond to the reality of the signal? It works by applying a twiddle factor to each stage of the calculation.

I purposely calculated the DFT for one of the even frequencies, $k=2$, to show how combining two 2-point DFTs helps us to calculate a 4-point DFT. However, if we try to do the same for one of the odd frequencies, $k=1$ for example, we hit a problem.

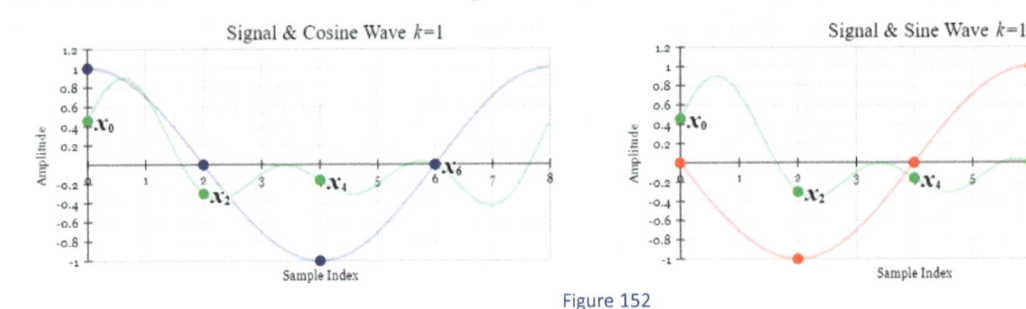

Figure 152

The problem with the FFT

If we use the long way round as before, we can see that the cosine component doesn't present a problem. Multiplying all the green points on the signal by their corresponding blue points on the cosine wave from Figure 153, we can derive Equation 58. The points on the cosine wave corresponding to points x_2 and x_6 on the signal are both zero, so multiplying the signal by the cosine wave at these points yields zero. This is why x_2 and x_6 do not appear in Equation 58.

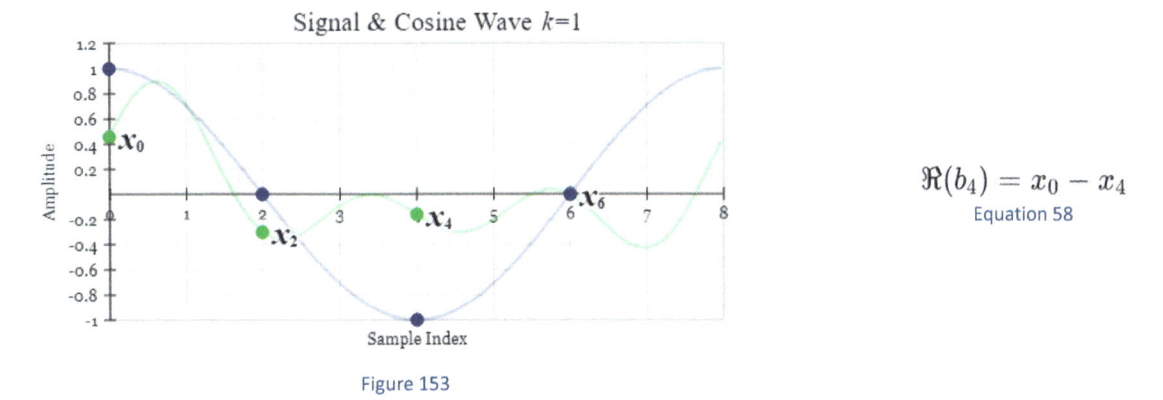

Signal & Cosine Wave $k=1$

$$\Re(b_4) = x_0 - x_4$$
Equation 58

Figure 153

However, we run into trouble when calculating the sine component, because, unlike all the 2-point DFTs from before, it is not zero. From Figure 154, we derive Equation 59. Here too, two of the points on the signal have been multiplied by points that are zero on the sine wave. This is why x_0 and x_4 do not appear in Equation 59.

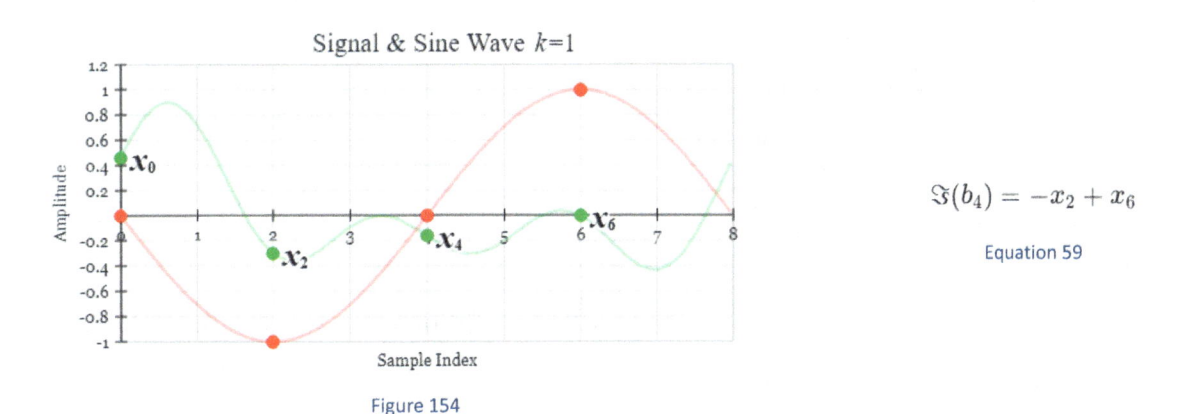

Signal & Sine Wave $k=1$

$$\Im(b_4) = -x_2 + x_6$$
Equation 59

Figure 154

When we combine the real part of b_4 from Equation 58 and the imaginary part of b_4 from Equation 59 using the DFT equation, we get Equation 60.

$$b_4 = \Re(b_4) - \Im(b_4)$$

$$= x_0 - x_4 - x_2 i + x_6 i$$
Equation 60

Notice that x_2 and x_6 in Equation 60 have both been multiplied by i. This is because they are imaginary, i.e., part of the sine component. The i makes sure that they cannot be added directly to x_0 and x_4.

We can group the real and imaginary parts of Equation 60 to get Equation 61.

$$b_4 = (x_0 - x_4) - (x_2 - x_6)i$$

Equation 61

Putting x_2 and x_6 inside brackets in Equation 61 changes their signs because of the subtraction between the real and imaginary parts of the equation. This is fortunate, as the terms inside the brackets in Equation 61 look familiar. We already did some of the work for this calculation during the 2-point DFT stage when we found that $(x_0 - x_4)$ was equal to a_4, and $(x_2 - x_6)$ was equal to a_6.

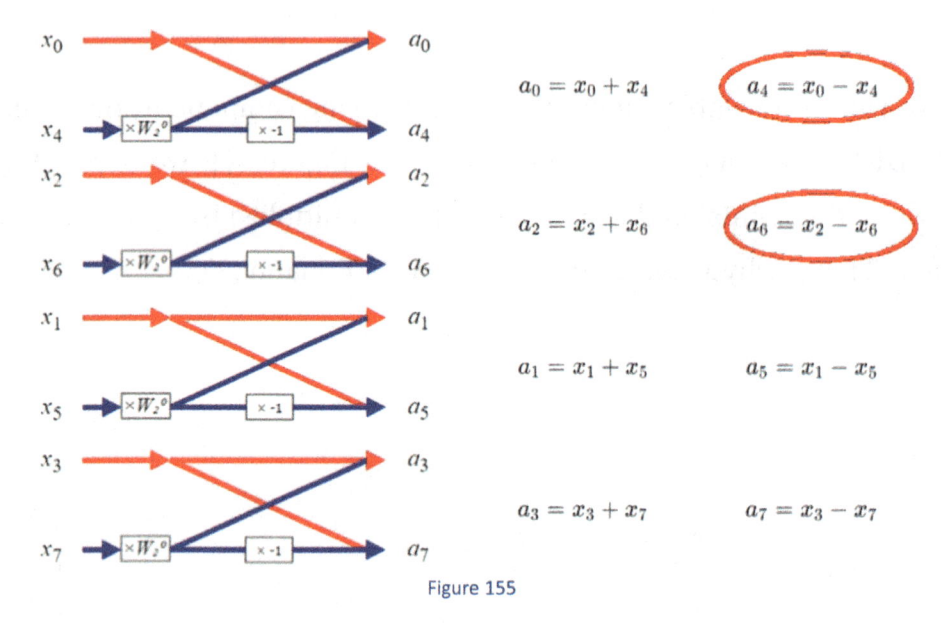

Figure 155

Therefore, we should be able to combine the results a_4 and a_6 from the 2-point DFTs, to calculate the 4-point DFT. We can represent this on a butterfly diagram like the one in Figure 156.

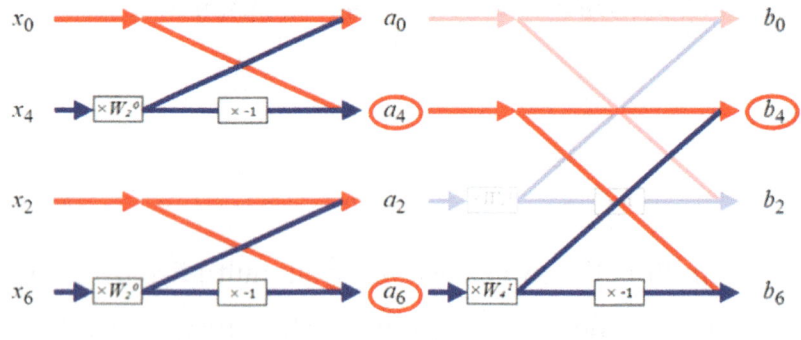

Figure 156

However, simply adding a_4 and a_6 together as the DFT requires will not give us the 4-point DFT for these samples, as we can see from Equation 62.

$$\frac{a_4}{(x_0 - x_4)} + \frac{a_6}{(x_2 - x_6)} \neq \frac{a_4}{(x_0 - x_4)} - \frac{a_6 i}{(x_2 - x_6)i}$$

Equation 62

We need to subtract them instead, and multiply a_6 by i. We cannot use a_6 directly. Does this mean we did something wrong when we calculated a_6 during the 2-point DFT stage? Yes, we did. We cheated!

In order to make things simple, we performed exactly the same calculation on each of the sample pairs.

Figure 157 demonstrates why that doesn't work. It shows samples x_2 and x_6 during the 2-point DFT stage. Notice that because of their position on the x-axis, they're not aligned with the sample points on the cosine wave for a 2-point DFT. This is a digital signal. It is defined only at discrete moments in time. In a 2-point DFT, the signal is defined only at points 0 and 4 on the x-axis, samples x_2 and x_6 fall in-between these points.

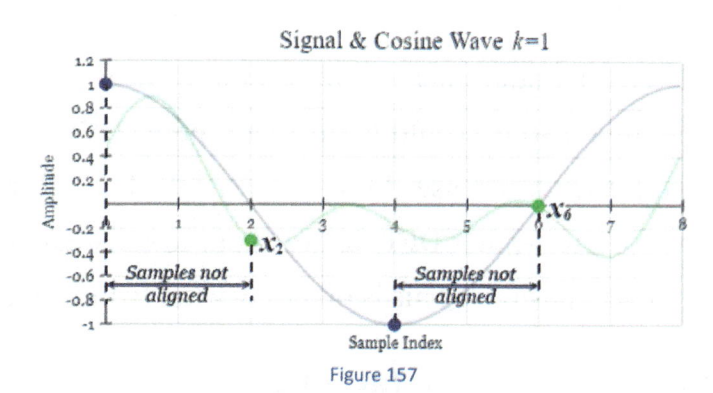

Figure 157

Returning to the spotlight analogy I used back in Chapter 1, it's as if samples x_2 and x_6 are not in the correct positions to be in the 2-point DFT's spotlight, and consequently it cannot see them as Figure 158 illustrates. So how did we perform a 2-point DFT on these samples?

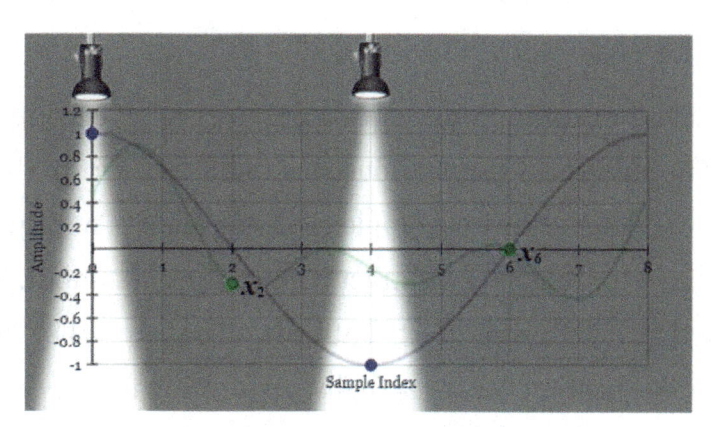

Figure 158

How the Fourier Transform works Chapter 5

We behaved as if the signal were shifted to the left along the x-axis, moving samples x_2 and x_6 into the light. This shifted them into alignment with the sample points on the cosine wave for a 2-point DFT, as shown in Figure 159.

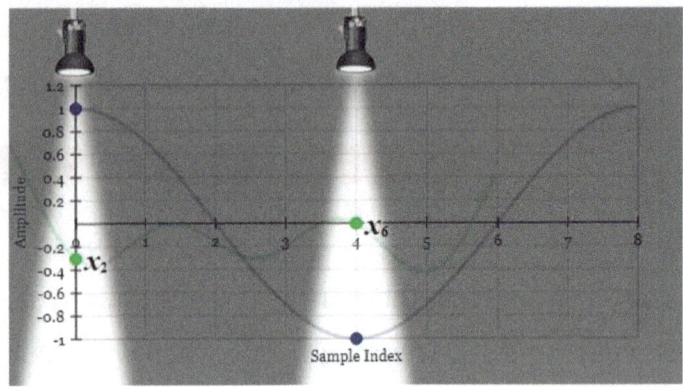

Figure 159

But surely we can't go around simply shifting our signal in any way we please just to make our sums easier to do! Won't this cause an error in the rest of the calculation? Of course it will. This is why we need a twiddle factor.

The verb "twiddle" means: "play around with." We have played around with the signal; we shifted it so we could perform a 2-point DFT on x_2 and x_6. Therefore, before we can move forward and use the result we calculated during the 2-point stage as a starting point for the 4-point DFT we are about to calculate, we need to shift the signal back into the correct position. This is precisely what twiddle factors do.

What are twiddle factors?

Twiddle factors are sometimes called phase factors. They change the phase of the signal, moving it along the x-axis. Twiddle factors are easy to spot. They are the blocks in the butterfly diagram which look something like Figure 160.

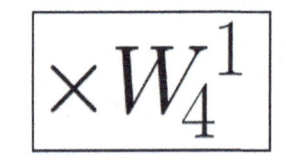

$$\times W_4^1$$

Figure 160

Each butterfly produces two outputs, an even output (red) and an odd output (blue). The outputs of the previous stage's butterfly form the inputs of the next stage's butterfly. A twiddle factor is a complex constant that is multiplied by the odd input, preparing it so that we can add or subtract it from the even input. This produces a new pair of outputs feeding into the next butterfly. The fact that it is complex means that in one multiplication we are actually performing two operations: a scaling operation (changing the input's amplitude), and a phase shift operation (shifting it along the x-axis).

https://howthefouriertransformworks.com/ Page 79

Let's look again at samples x_2 and x_6 as they flow through the 2-point butterfly and into the 4-point butterfly.

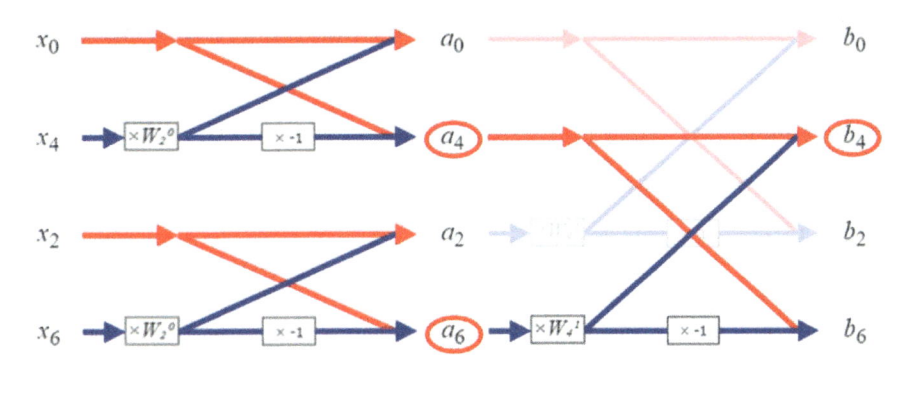

Figure 161

Sample x_6 enters the 2-point butterfly at the bottom left of Figure 161. It is then multiplied by the twiddle factor W_2^0. As we'll find out shortly, this twiddle factor is equal to 1, so x_6 remains unchanged. It is then multiplied by -1 and added to x_2 to produce a_6. In effect, all we have done to produce a_6 is to subtract x_6 from x_2.

Value a_6 is the odd output from this 2-point butterfly. This output then forms one of the odd inputs to the 4-point butterfly. Remember, a 4-point butterfly is simply two interleaved 2-point butterflies. We multiply a_6 by the twiddle factor W_4^1. This scales a_6 and also changes its phase. The change in phase has effectively shifted the samples making up a_6 along the x-axis, into the correct position for a 4-point DFT. The scaling operation has scaled the amplitude of a_6, so that it is as though the samples making it up have been multiplied by the corresponding samples on a sine wave.

Calculating twiddle factors

Let's find out why multiplying the odd input of the 2-point butterfly by the twiddle factor W_2^0 is the same as multiplying by 1.

Twiddle Factor: W_2^0

The notation for a twiddle factor provides a clue for how to calculate it.

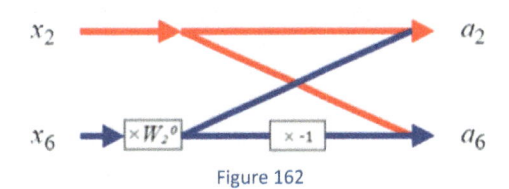

Figure 162

The subscript number tells us the order of the DFTs we are currently working with. Here the order is 2, as we are working with a 2-point DFT.

The superscript number tells us the index of the particular sample within the current DFT. Since only the odd (blue) inputs are multiplied by twiddle factors, the number of possible index values is only half the value of the order. In this example, the order is 2, so there is only one index value. And since the index value always starts with 0, the index in a 2-point butterfly will always be 0.

To calculate the twiddle factor, we use Equation 63, where O is the order, and I is the index:

$$W_O^I = cos\left(2\pi \cdot \frac{I}{O}\right) - i \cdot sin\left(2\pi \cdot \frac{I}{O}\right)$$

Equation 63

Equation 64 shows the calculation for twiddle factor W_2^0.

$$W_2^0 = cos\left(2\pi \cdot \frac{0}{2}\right) - i \cdot sin\left(2\pi \cdot \frac{0}{2}\right)$$

Equation 64

Equation 64 simplifies to Equation 65.

$$W_2^0 = cos(0) - i \cdot sin(0)$$

Equation 65

Since $cos(0) = 1$ and $sin(0) = 0$, we can calculate Equation 66.

$$W_2^0 = 1 - 0i = 1$$

Equation 66

Notice that there is no imaginary part in Equation 66, as i is multiplied by zero. The real part is equal to 1. Therefore, the twiddle factor W_2^0 is equal to 1. The same is true for any twiddle factor with an index of zero, because the index, I, is in the numerator of Equation 63. If the numerator is zero, the entire expression inside the brackets will be zero.

The absence of an imaginary part to the twiddle factor means that there is no phase shift, and this operation becomes a straight forward scaling operation. However, the scaling operation is a multiplication by 1. All this means that twiddle factor W_2^0 actually has no effect on the signal. It is included in the butterfly simply to ensure that the notation for all the butterflies in the diagram remains consistent.

Twiddle Factor: W_4^1

Looking at the butterfly diagram for the 4-point butterfly in Figure 163, the input a_6 is multiplied by the twiddle factor W_4^1. The subscript number tells us that the order of the DFTs we are currently working with is 4.

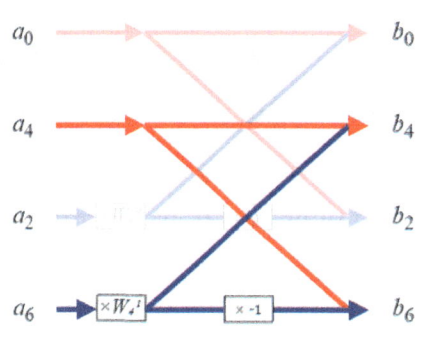

Figure 163

The superscript number tells us the index of the particular sample within the current DFT. Since only the odd (blue) inputs are multiplied by twiddle factors, the number of possible index values is only half the value of the order. In this example, the order is 4, so there are two index values, 0 and 1. As this is the odd input of the second of the interleaved butterflies, the index is 1.

Twiddle factor W_4^1 can thus be calculated as shown in Equation 67.

$$W_4^1 = \cos\left(2\pi \cdot \frac{1}{4}\right) - i \cdot \sin\left(2\pi \cdot \frac{1}{4}\right)$$

Equation 67

Equation 67 simplifies to Equation 68,

$$W_4^1 = \cos\left(\frac{\pi}{2}\right) - i \cdot \sin\left(\frac{\pi}{2}\right)$$

Equation 68

which yields Equation 69.

$$W_4^1 = -i$$

Equation 69

The presence of an imaginary part means that multiplying by the twiddle factor W_4^1 has shifted the phase and scaled the amplitude of a_6. The samples x_2 and x_6, which make up a_6, have effectively been moved from the positions they occupied during the 2-point DFT.

They've been shifted along the x-axis by -90°, into the position now necessary for the 4-point DFT, and the amplitude of a_6 has been scaled by -1. We can see this in Figure 164.

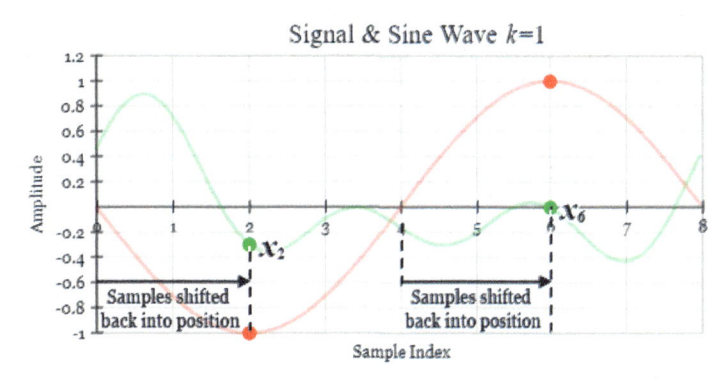

Equation 70 shows mathematically how the twiddle factor, W_4^1, allows us to calculate b_4 using a_4 and a_6. W_4^1 has shifted the phase of a_6, and Equation 70 is now equal to Equation 61.

$$\begin{aligned} b_4 &= a_4 + a_6 \times W_4^1 \\ &= a_4 + a_6 \times -i \\ &= a_4 - a_6 i \\ &= (x_0 - x_4) - (x_2 - x_6)i \end{aligned}$$

Figure 164

Equation 70

A multiplication by $-i$ is a rotation of $-90°$ ($-\pi/2$ radians) into the imaginary dimension. In our terms, that means a $-90°$ phase shift along the x-axis, which is exactly what has happened to a_6.

For a video explaining why multiplying by i causes a rotation or phase shift, scan the QR code or click on the following link:

The imaginary number i and the Fourier Transform

Calculating the second group of 4-point DFTs

Now that we've calculated the first group of 4-point DFTs, involving a_0, a_2, a_4 and a_6, how do we calculate a_1, a_3, a_5, and a_7? They are located in the second group of 4-point butterflies. We calculate this second group in EXACTLY the same way as we calculated the first group.

Note that the word EXACTLY is capitalized. This is the whole point of the FFT's use of the divide-and-conquer algorithm. Every stage of the process is treated in exactly the same way. Remember, this will cause a phase problem as we move into the final stage of the calculation, the 8-point butterfly, which we'll have to fix with more twiddle factors, as shown in Figure 165. However, that's a problem we'll deal with later. As far as this stage of the FFT is concerned, we treat a_1, a_3, a_5, and a_7 in exactly the same way as we treated a_0, a_2, a_4, and a_6.

How the Fourier Transform works　　　　　　　　　　　　　　　　　Chapter 5

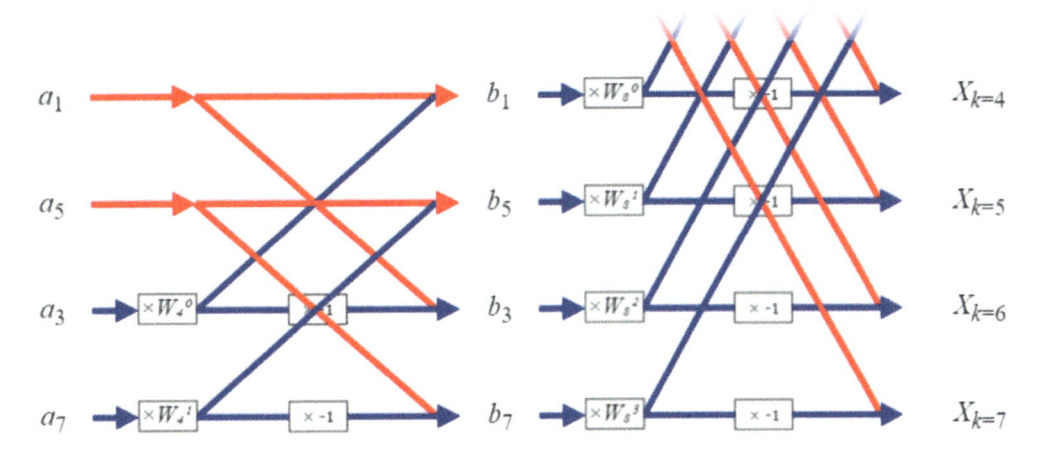

(Figure 165: butterfly diagram showing inputs a_1, a_5, a_3, a_7 with twiddle factors $\times W_4^0$, $\times W_4^1$, combining to b_1, b_5, b_3, b_7, then through $\times W_8^0, \times W_8^1, \times W_8^2, \times W_8^3$ and $\times -1$ stages to outputs $X_{k=4}, X_{k=5}, X_{k=6}, X_{k=7}$.)

Figure 165

Figure 166 reminds us why the twiddle factors are necessary. The samples x_1, x_3, x_5, and x_7 on the signal are not aligned with the sample points on the test wave for a 4-point DFT.

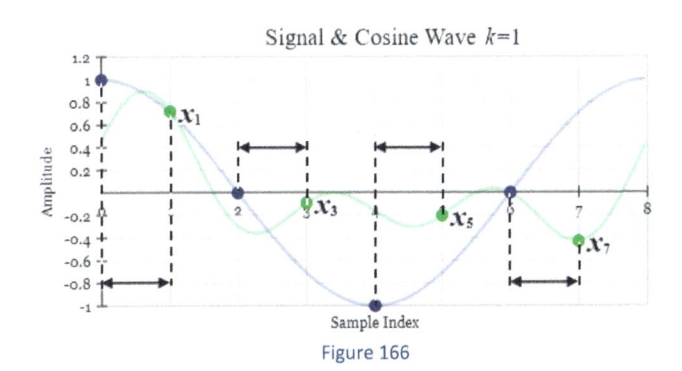

(Figure 166: Signal & Cosine Wave $k=1$, showing x_1, x_3, x_5, x_7 plotted against sample index.)

Figure 166

The test wave shown here is the cosine component at frequency index $k=1$. However, the same is true for all the test waves, both cosine and sine, at all the frequency indexes tested in a 4-point DFT ($k=0$, $k=1$, $k=2$, and $k=3$).

We then perform the calculation exactly as we did for the first group of 4-point DFTs, building on the results we already calculated in the 2-point DFT stage.

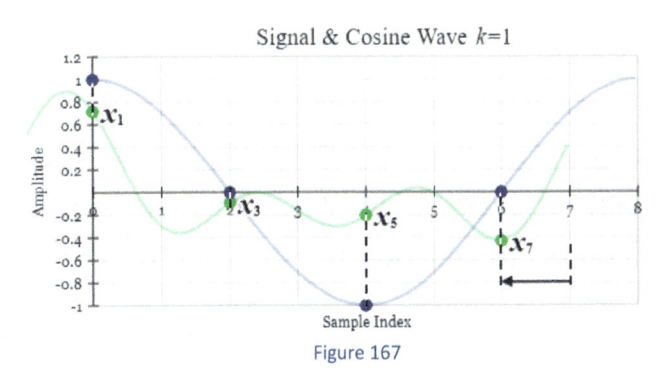

(Figure 167: Signal & Cosine Wave $k=1$, showing x_1, x_3, x_5, x_7 plotted against sample index.)

Figure 167

https://howthefouriertransformworks.com/　　　　　　　　　　　　　　　　Page 84

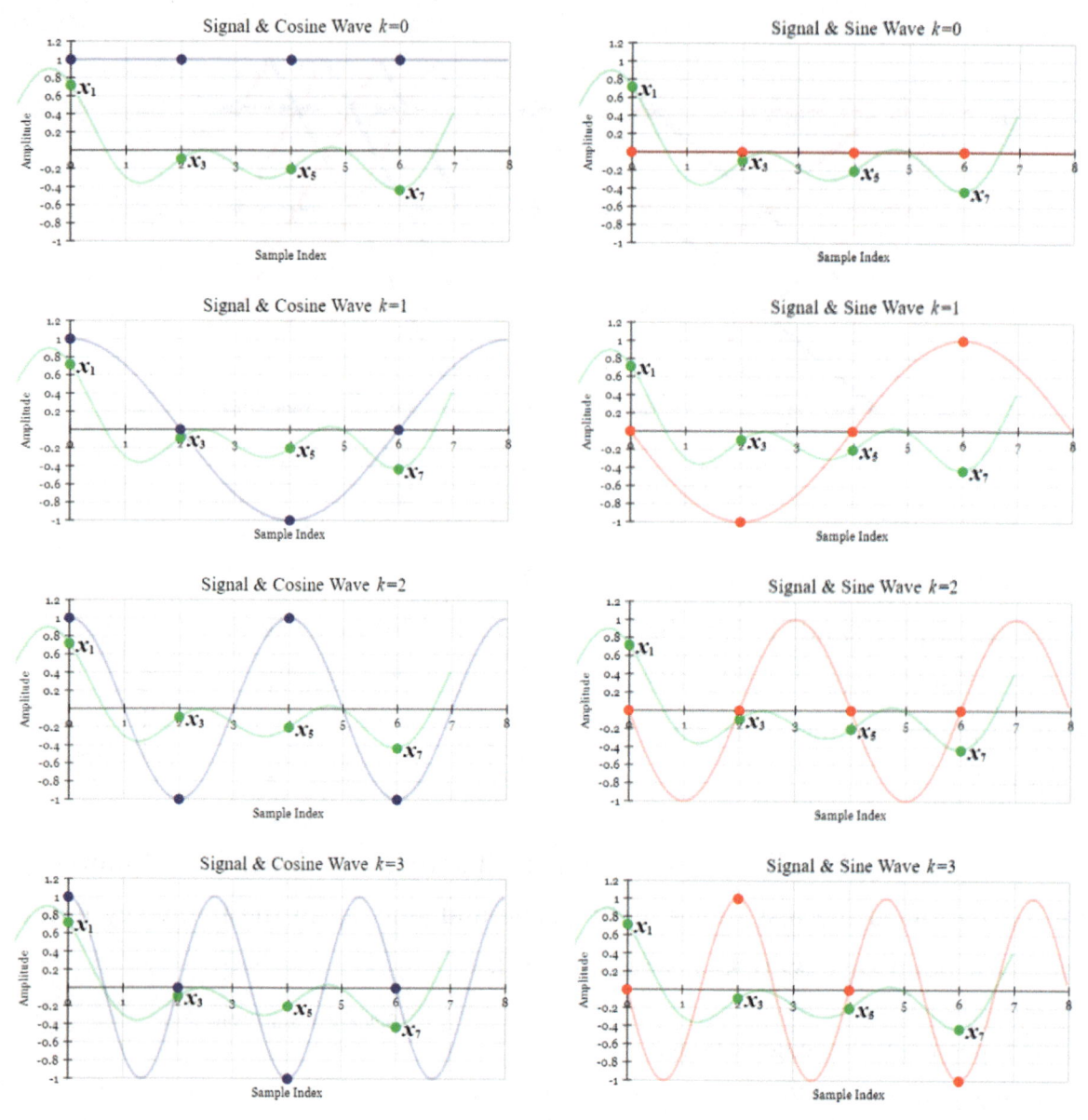

Figure 168

During the 2-point DFT stage we already calculated a_1, a_3, a_5, and a_7, as shown in Equation 71 to Equation 74.

$$a_1 = x_1 + x_5$$

Equation 71

$$a_3 = x_3 + x_7$$

Equation 72

$$a_5 = x_1 - x_5$$

Equation 73

$$a_7 = x_3 - x_7$$

Equation 74

Now we can use the upper butterfly in the second group of 4-Point DFTs to calculate b_1 and b_3.

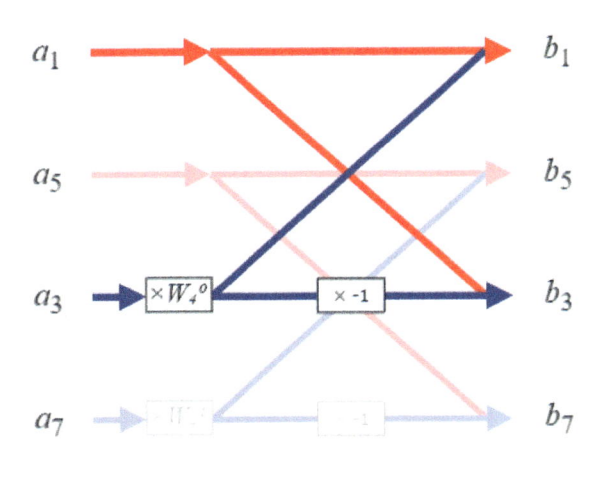

Figure 169

$$b_1 = a_1 + W_4^0 \times a_3$$

Equation 75

$$b_3 = a_1 - W_4^0 \times a_3$$

Equation 76

In the section on calculating twiddle factors, we discovered any twiddle factor with an index of 0, like W_4^0, is equal to 1. Therefore we can rewrite the calculation for b_1 and b_3, as shown in Equation 77 and Equation 78.

$$b_1 = a_1 + a_3$$

Equation 77

$$b_3 = a_1 - a_3$$

Equation 78

We can then use the lower butterfly in the second group of 4-point DFTs to calculate b_5 and b_7.

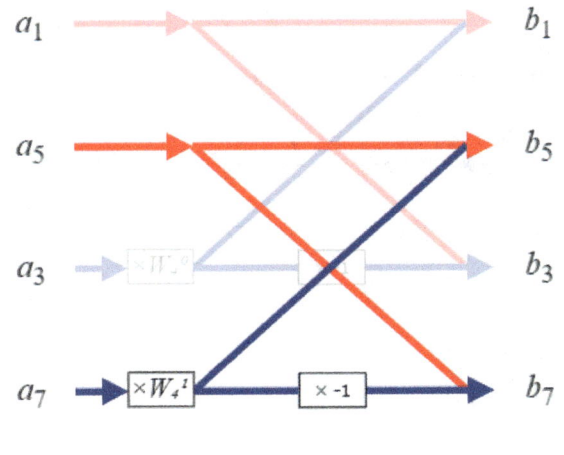

Figure 170

$$b_5 = a_5 + W_4^1 \times a_7$$

Equation 79

$$b_7 = a_5 - W_4^1 \times a_7$$

Equation 80

In the section on calculating twiddle factors, we discovered the twiddle factor W_4^1, is equal to $-i$, so we can rewrite the calculations for b_5 and b_7 as shown in Equation 81 and Equation 82.

$$b_5 = a_5 - a_7 \cdot i$$

Equation 81

$$b_7 = a_5 + a_7 \cdot i$$

Equation 82

As long as our signal contains a number of samples that is a power of 2, the method for calculating each stage of the FFT remains the same. Our signal contains eight samples, so we can perform an 8-point FFT on it. So far, we've covered the 2-point and 4-point DFT stages. The 8-point DFT stage, which is the final stage for this signal, is calculated in exactly the same way.

The full butterfly diagram for an 8-point FFT is shown in Figure 171.

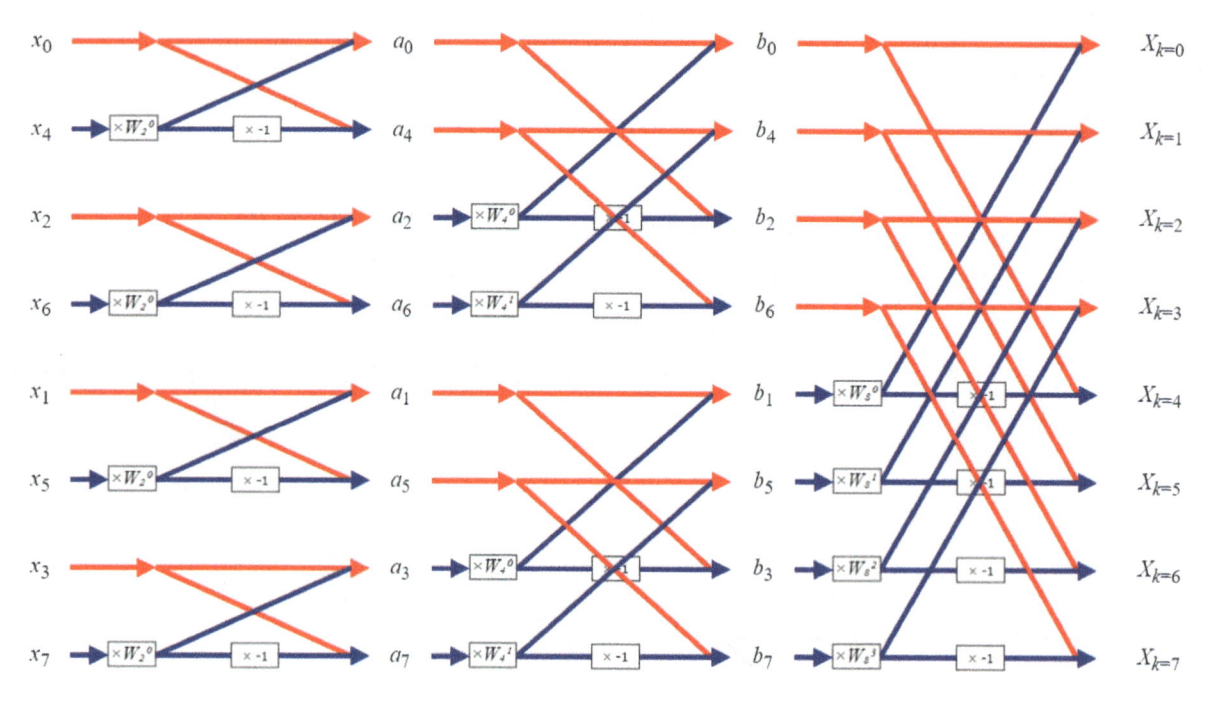

Figure 171

Calculating the FFT – a numerical example

Let us take all that we've learned, and calculate the FFT for the signal shown in Figure 172. I've listed the sample values for the signal in Figure 173.

Figure 172

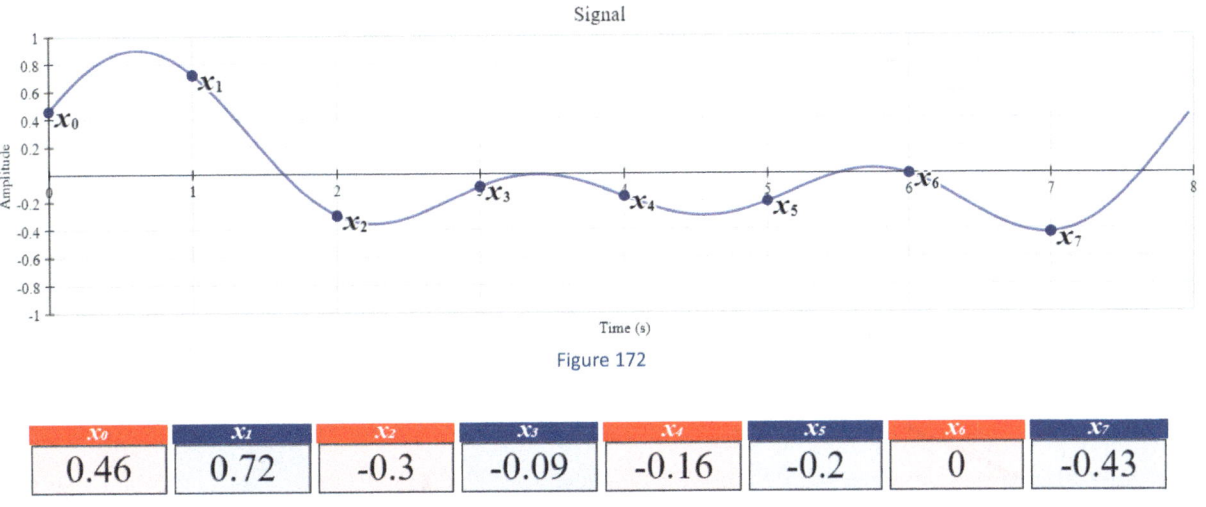

x_0	x_1	x_2	x_3	x_4	x_5	x_6	x_7
0.46	0.72	-0.3	-0.09	-0.16	-0.2	0	-0.43

Figure 173

Calculating the DFT of frequency index $k=0$

The graphs in Figure 174 show the signal, cosine test wave, and sine test wave taking part in the calculation.

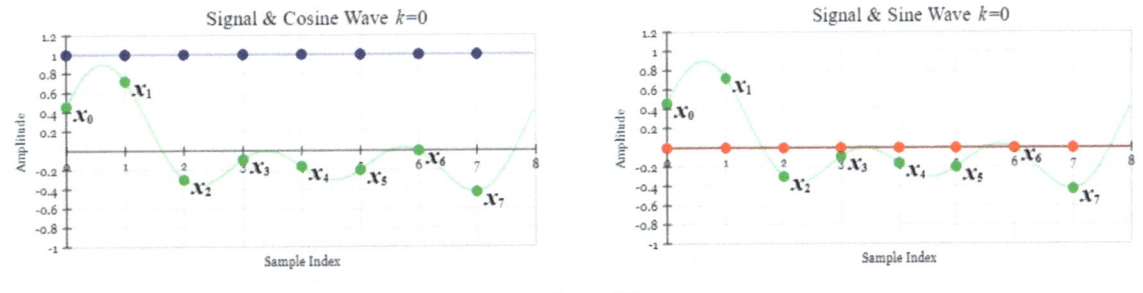

Figure 174

The sample values, x_0 to x_7, are taken from the signal (the green points on the graphs above). We can map the path these samples take through the FFT algorithm on the butterfly diagram in Figure 175, all the way to the output, $X_{k=0}$, for frequency index $k=0$.

Figure 175

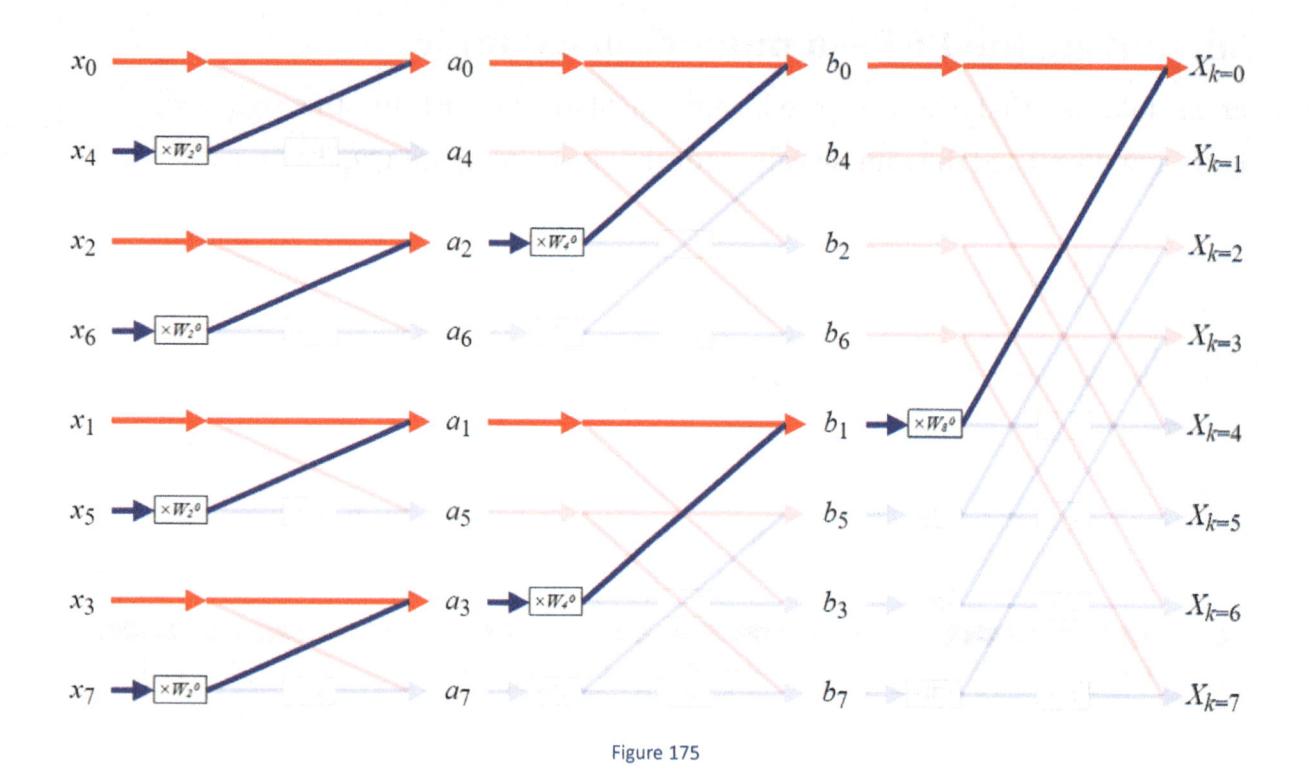

Twiddle Factor W_2^0

Although samples x_1, x_3, x_5, and x_7 all pass through the twiddle factor W_2^0 on the butterfly diagram, I haven't bothered to write it in any of the following calculations. The reason is that the index of the twiddle factor is 0, so $W_2^0 = 1$, as shown in Equation 83.

Therefore, multiplying anything by this twiddle factor has no effect. I have included it in the butterfly diagram only to keep the notation consistent with the way the FFT runs its calculations.

$$\begin{aligned} W_2^0 &= cos\left(2\pi \cdot \tfrac{0}{2}\right) - i \cdot sin\left(2\pi \cdot \tfrac{0}{2}\right) \\ &= cos(0) - i \cdot sin(0) \\ &= 1 \end{aligned}$$

Equation 83

How the Fourier Transform works — Chapter 5

The equation to calculate the DFT at this frequency

From the butterfly diagram in Figure 175, we can derive Equation 84.

$$X_{k=0} = \left[(x_0 + x_4) + W_4^0 \cdot (x_2 + x_6)\right] + W_8^0 \cdot \left[(x_1 + x_5) + W_4^0 \cdot (x_3 + x_7)\right]$$

$$= (a_0 + W_4^0 \cdot a_2) + W_8^0 \cdot (a_1 + W_4^0 \cdot a_3)$$

$$= b_0 + W_8^0 \cdot b_1$$

Equation 84

The green blocks are the samples x_0 to x_7, which form the inputs to the 2-point DFT stage. The blue numbers in each block are the actual sample values. The samples are added together to produce the inputs to the 4-point DFT stage in the red blocks, a_0 to a_3, which are calculated according to Equation 85 to Equation 88.

$$a_0 = x_0 + x_4 = 0.46 + -0.16 = 0.30$$

Equation 85

$$a_2 = x_2 + x_6 = -0.30 + 0.00 = -0.30$$

Equation 86

$$a_1 = x_1 + x_5 = 0.72 + -0.20 = 0.52$$

Equation 87

$$a_3 = x_3 + x_7 = -0.09 + -0.43 = -0.52$$

Equation 88

Twiddle Factor W_4^0

Values a_2 and a_3 in Equation 84 are both multiplied by the twiddle factor W_4^0, as shown in Equation 89. The index of the twiddle factor is 0, so $W_4^0 = 1$.

$$W_4^0 = \cos\left(2\pi \cdot \tfrac{0}{4}\right) - i \cdot \sin\left(2\pi \cdot \tfrac{0}{4}\right)$$
$$= \cos(0) - i \cdot \sin(0)$$
$$= 1$$

Equation 89

Therefore, a_0 and a_2 can be added together directly to produce b_0, as shown in Equation 90. The same is true of a_1 and a_3, producing b_1 as shown in Equation 91. Now, b_0 and b_1 are the inputs to the 8-point DFT stage.

$$b_0 = a_0 + W_4^0 \times a_2 = 0.30 + 1 \times -0.30 = 0.00$$

Equation 90

$$b_1 = a_1 + W_4^0 \times a_3 = 0.52 + 1 \times -0.52 = 0.00$$

Equation 91

Twiddle Factor W_8^0

As we can see from Equation 84, b_1 is multiplied by the twiddle factor W_8^0. Again, the index of the twiddle factor is 0, meaning $W_8^0 = 1$.

Completing the calculation

The inputs b_0 and b_1 can therefore be added together directly to produce the result for this first frequency index $k=0$, as shown in Equation 92.

$$X_{k=0} = b_0 + W_8^0 \times b_1$$
$$= 0.00 + 1 \times 0.00$$
$$X_{k=0} = 0.00$$

Equation 92

Calculating the DFT of frequency index $k=4$

The beauty of the FFT is that having already calculated the DFT for one frequency, we get another frequency almost for free.

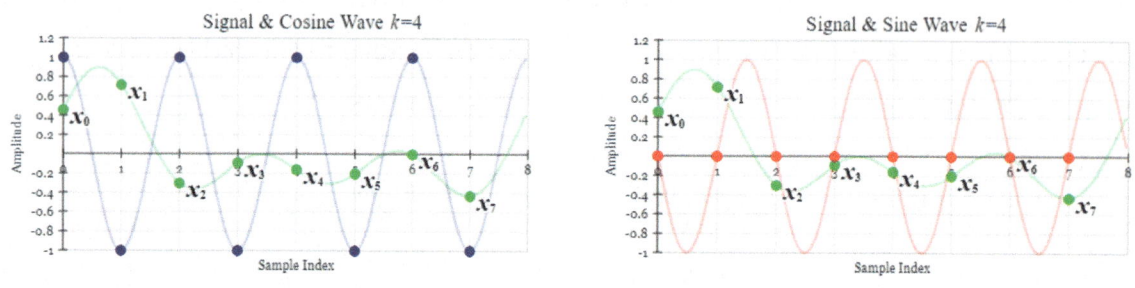

Figure 176

Because of the position of the samples, the only difference between the DFT at frequency index $k=0$ and the DFT at frequency index $k=4$ is that on the cosine graph, the left-hand graph in Figure 176, every other sample is a negative version of the sample before it.

On the sine graph, the right-hand graph in Figure 176, all the samples are zero, just as they were at frequency index $k=0$.

The DFT for frequency index $k=4$ can therefore be computed using almost the same butterfly diagram as we used to calculate the DFT for frequency index $k=0$. The only difference is we use the result from the lower branch of the final butterfly in the 8-point DFT stage, instead of the upper branch, as we can see in Figure 177.

How the Fourier Transform works   Chapter 5

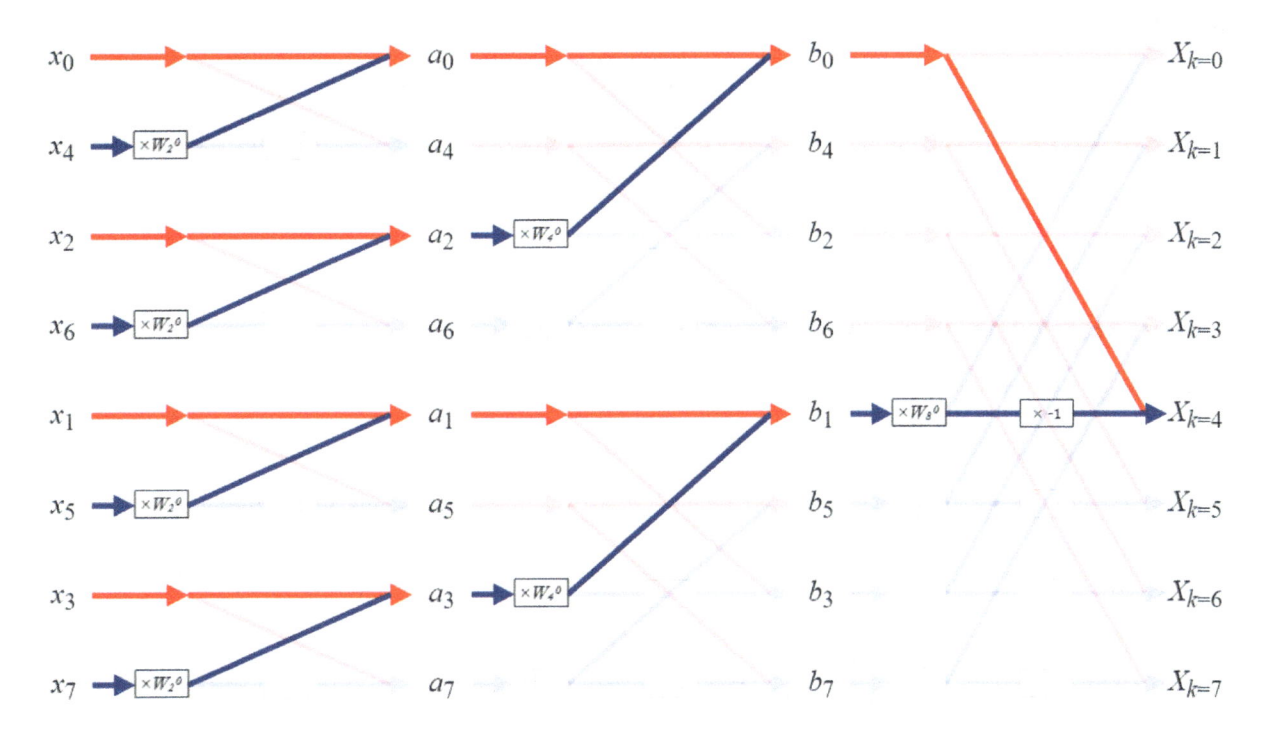

Figure 177

We can thus easily calculate the DFT for this frequency by making only one minor change to the equation, as shown in Equation 93. Instead of adding b_1 to b_0, we subtract it.

$$X_{k=4} = \left[(\overset{0.46}{x_0} + \overset{-0.16}{x_4}) + W_4^0 \cdot (\overset{-0.30}{x_2} + \overset{0.00}{x_6}) \right] - W_8^0 \cdot \left[(\overset{0.72}{x_1} + \overset{-0.20}{x_5}) + W_4^0 \cdot (\overset{-0.09}{x_3} + \overset{-0.43}{x_7}) \right]$$

$$= \left(\overset{0.30}{a_0} + W_4^0 \cdot \overset{-0.30}{a_2} \right) - W_8^0 \cdot \left(\overset{0.52}{a_1} + W_4^0 \cdot \overset{-0.52}{a_3} \right)$$

$$= \overset{0.00}{b_0} - W_8^0 \cdot \overset{0.00}{b_1}$$

Equation 93

Completing the calculation

The twiddle factors are the same as before, so we can disregard them, as they are both equal to 1. Therefore, calculating the DFT for this frequency is simplicity itself, as shown in Equation 94.

$$X_{k=4} = b_0 - W_8^0 \times b_1$$
$$= 0.00 - 1 \times 0.00$$
$$X_{k=4} = 0.00$$

Equation 94

Calculating the DFT of frequency index *k*=1

Now things get interesting. When dealing with the 8-point DFT for frequency index *k*=1, the amplitude of the samples on the cosine and sine waves are no longer only 1 or 0.

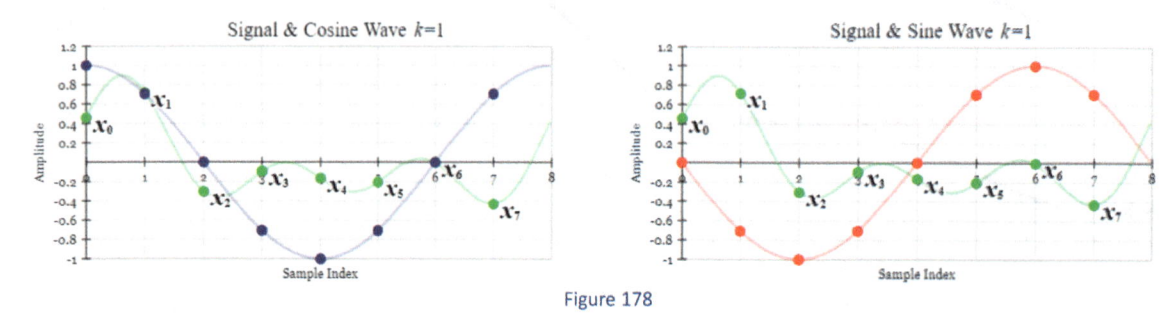

Figure 178

This is where the twiddle factors have an effect. Let's look at the butterfly diagram in Figure 179.

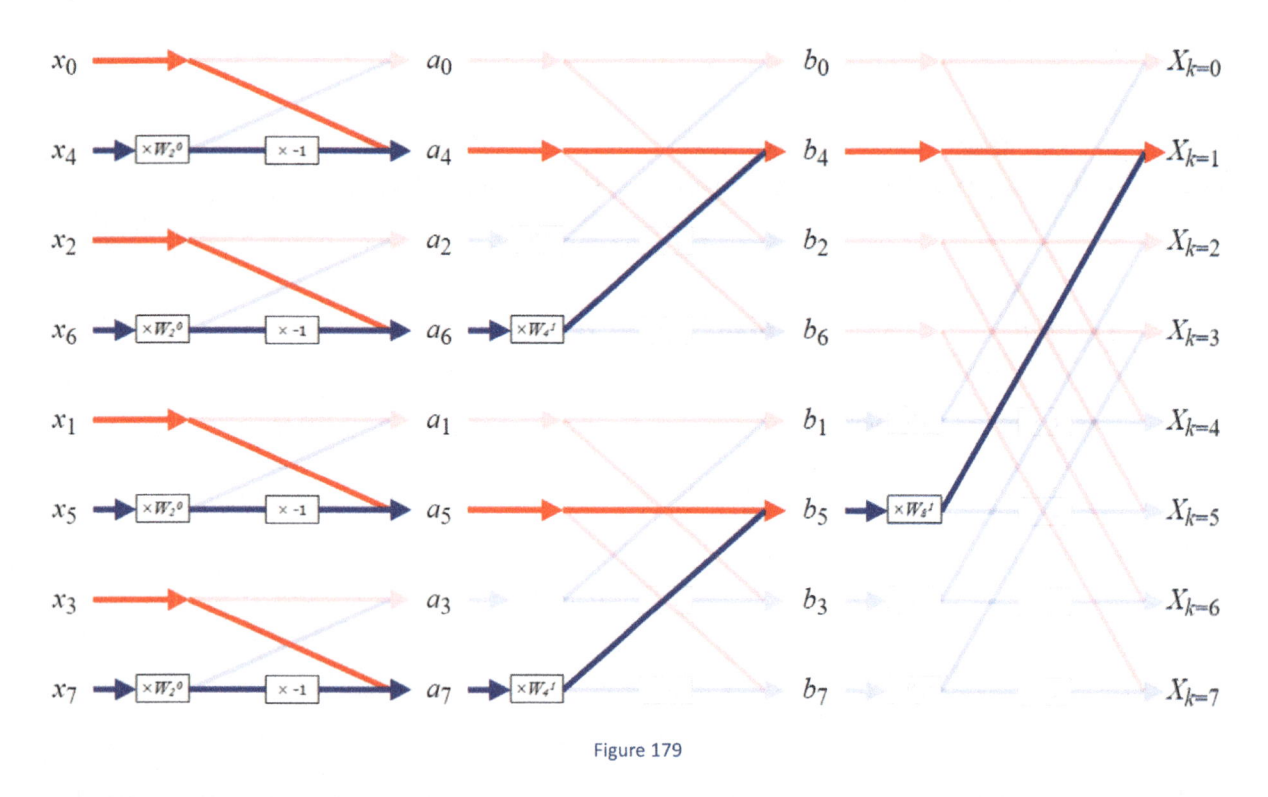

Figure 179

From the butterfly diagram, we can derive Equation 95.

$$X_{k=1} = \left[(x_0 - x_4) + W_4^1 \cdot (x_2 - x_6) \right] + W_8^1 \cdot \left[(x_1 - x_5) + W_4^1 \cdot (x_3 - x_7) \right]$$

$$= (a_4 + W_4^1 \cdot a_6) + W_8^1 \cdot (a_5 + W_4^1 \cdot a_7)$$

$$= b_4 + W_8^1 \cdot b_5$$

Equation 95

There are two twiddle factors in play for this frequency, W_4^1 and W_8^1.

Twiddle Factor W_4^1

As we have seen previously, we can calculate W_4^1 using Equation 96.

$$W_4^1 = \cos(2\pi \cdot \tfrac{1}{4}) - i \cdot \sin(2\pi \cdot \tfrac{1}{4})$$
$$= \cos(\tfrac{\pi}{2}) - i \cdot \sin(\tfrac{\pi}{2})$$
$$= -i$$

Equation 96

We can then calculate b_4 and b_5 using Equation 97 and Equation 98.

$$b_4 = a_4 + W_4^1 \times a_6$$
$$= 0.62 + -i \times -0.30$$
$$= 0.62 + 0.30i$$

Equation 97

$$b_5 = a_5 + W_4^1 \times a_7$$
$$= 0.92 + -i \times 0.34$$
$$= 0.92 - 0.34i$$

Equation 98

Twiddle Factor W_8^1

We can calculate W_8^1 using Equation 99.

$$W_8^1 = \cos(2\pi \cdot \tfrac{1}{8}) - i \cdot \sin(2\pi \cdot \tfrac{1}{8})$$
$$= \cos(\tfrac{\pi}{4}) - i \cdot \sin(\tfrac{\pi}{4})$$
$$= 0.707 - 0.707i$$

Equation 99

The value b_5 is multiplied by the twiddle factor W_8^1, so the DFT for frequency index $k=1$ can be calculated using Equation 100.

$$X_{k=1} = b_4 + W_8^1 \times b_5$$
$$= (0.62 + 0.30i) + (0.707 - 0.707i) \times (0.92 - 0.34i)$$

Equation 100

We now have to perform two complex calculations to solve the above equation: first a complex multiplication, to find the product, $W_8^1 \times b_5$, and then a complex addition, to add the result of the product to b_4.

Complex Multiplication

The method used for complex multiplication is the same as the method for multiplying two brackets: the FOIL method. FOIL stands for First, Outside, Inside, Last. The multiplication is performed in four separate stages.

> **First:** We multiply the first terms in each bracket.
>
> **Outside:** We multiply the outside terms of the two brackets.
>
> **Inside:** We multiply the inside terms of the two brackets.
>
> **Last:** We multiply the last terms in each bracket.

Equation 101 shows how the FOIL method is used to perform the complex multiplication on W_8^1 and b_5.

$$
\begin{aligned}
W_8^1 \times b_5 = & \; (0.707 - 0.707i) \times (0.92 - 0.34i) \\
& \boxed{\begin{aligned}
\text{First:} & \quad 0.707 \times 0.92 = 0.65 \\
\text{Outside:} & \quad 0.707 \times -0.34i = -0.24\,i \\
\text{Inside:} & \quad -0.707i \times 0.92 = -0.65\,i \\
\text{Last:} & \quad -0.707i \times -0.34i = 0.24\,i^2 \\
& \qquad\qquad\qquad\qquad\quad = 0.24 \times -1 \\
& \qquad\qquad\qquad\qquad\quad = -0.24
\end{aligned}} \\
W_8^1 \times b_5 = & \; 0.65 - 0.24i - 0.65i - 0.24 \\
= & \; 0.65 - 0.24 - 0.24i - 0.65i \\
= & \; 0.41 - 0.89i
\end{aligned}
$$

Equation 101

Completing the calculation with complex addition

We add two complex numbers by grouping together all the real terms in one group, and all the imaginary terms in another, as shown in Equation 102.

$$
\begin{aligned}
X_{k=1} &= b_4 + W_8^1 \times b_5 \\
&= (0.62 + 0.3i) + (0.41 - 0.89i) \\
&= 0.62 + 0.41 + 0.3i - 0.89i \\
&= \boxed{X_{k=1} = 1.03 - 0.59i}
\end{aligned}
$$

Equation 102

Calculating the DFT of frequency index *k*=5

Once again, having calculated the DFT for frequency index *k*=1, we get the DFT of frequency index *k*=5 almost for free.

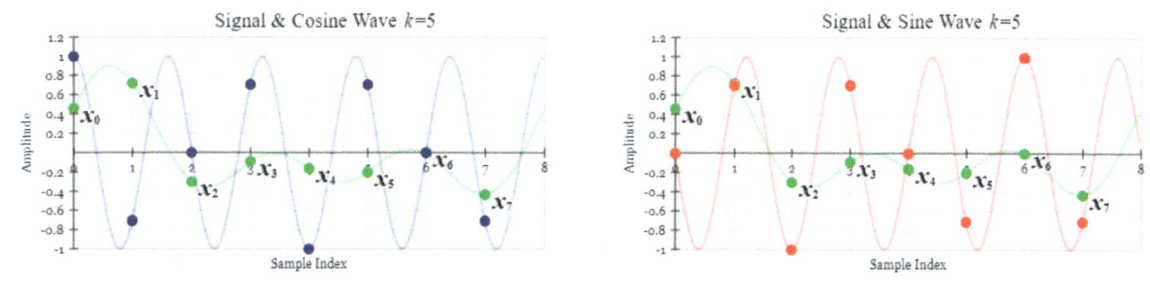

Figure 180

The DFT for frequency index *k*=5 can be computed using almost the same butterfly diagram as we used to calculate the DFT for frequency index *k*=1. The only difference is that we use the result from the lower branch of the final butterfly in the 8-point DFT stage, instead of the upper branch.

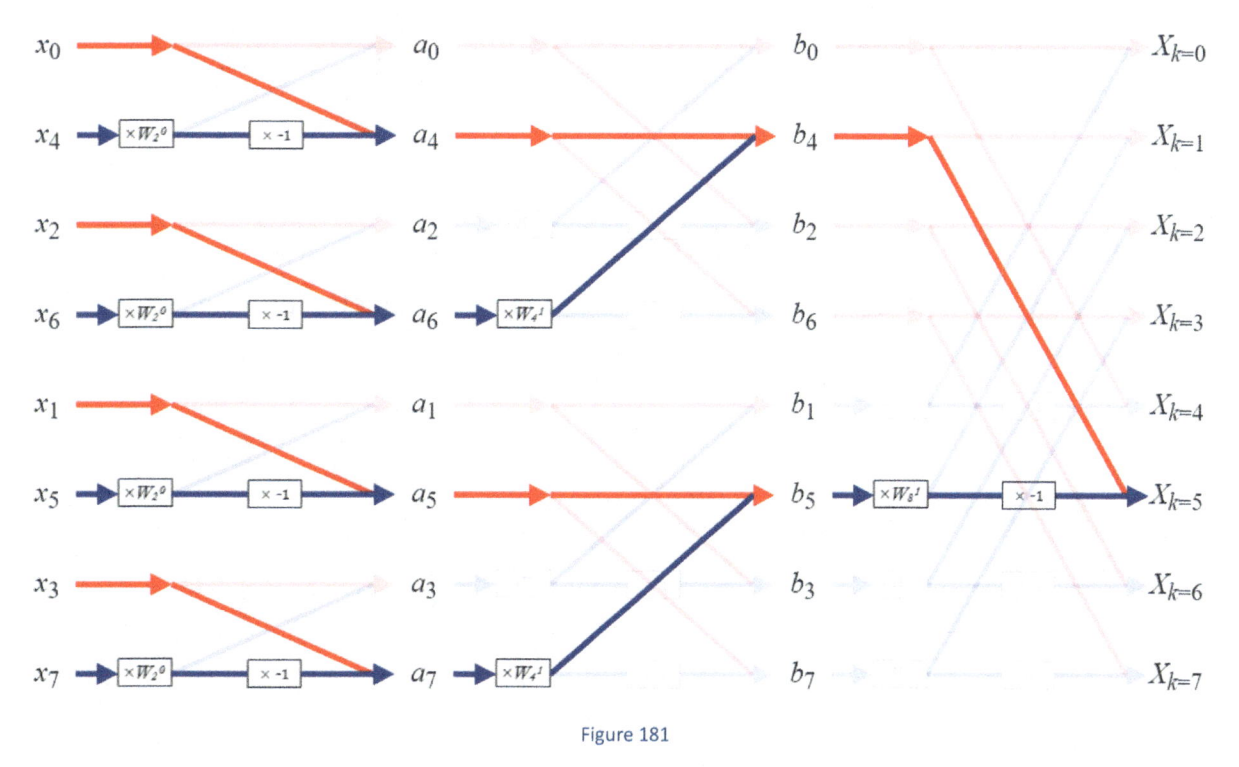

Figure 181

We can therefore calculate the DFT for this frequency easily, by making only one minor change to Equation 95. Instead of adding b_5 to b_4, we subtract it.

$$X_{k=5} = \left[(x_0 - x_4) + W_4^1 \cdot (x_2 - x_6) \right] - W_8^1 \cdot \left[(x_1 - x_5) + W_4^1 \cdot (x_3 - x_7) \right]$$

$$= (a_4 + W_4^1 \cdot a_6) - W_8^1 \cdot (a_5 + W_4^1 \cdot a_7)$$

$$= b_4 - W_8^1 \cdot b_5$$

Equation 103

Completing the calculation

Our complex addition from frequency index $k=1$ becomes a complex subtraction for frequency index $k=5$.

Subtracting two complex numbers is like adding them; we must first group together all the real terms in one group, and all the imaginary terms in another, as shown in Equation 104.

$$\begin{aligned} X_{k=5} &= b_4 - W_8^1 \times b_5 \\ &= (0.62 + 0.3i) - (0.41 - 0.89i) \\ &= 0.62 - 0.41 + 0.3i + 0.89i \\ &= \boxed{X_{k=5} = 0.21 + 1.19i} \end{aligned}$$

Equation 104

Calculating the DFT of frequency index $k=2$

We have now learned all the basic skills needed to complete the rest of the FFT algorithm. We need only to reuse those skills to complete the calculation.

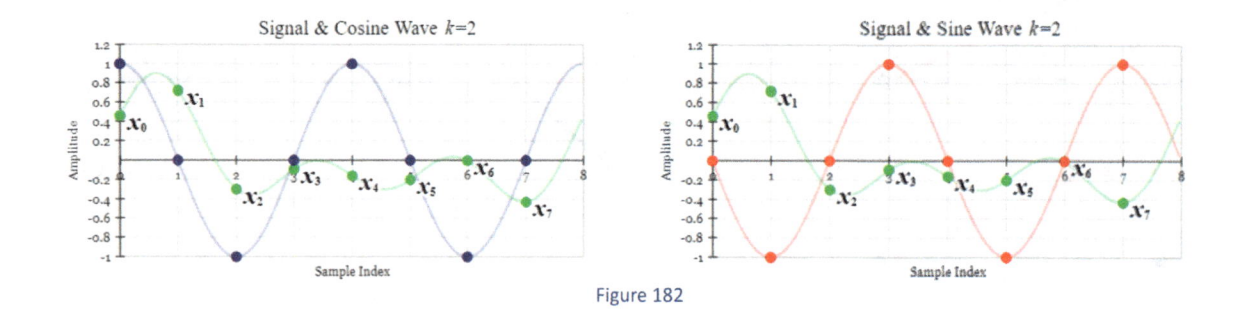

Figure 182

How the Fourier Transform works — Chapter 5

Figure 183

$$X_{k=2} = \left[(x_0 + x_4) - W_4^0 \cdot (x_2 + x_6) \right] + W_8^2 \cdot \left[(x_1 + x_5) - W_4^0 \cdot (x_3 + x_7) \right]$$

$$= \left(a_0 - W_4^0 \cdot a_2 \right) + W_8^2 \cdot \left(a_1 - W_4^0 \cdot a_3 \right)$$

$$= b_2 + W_8^2 \cdot b_3$$

With values:
- $x_0 + x_4 = 0.46$, $x_2 + x_6 = -0.16$, (etc. -0.30, 0.00, 0.72, -0.20, -0.09, -0.43)
- $a_0 = 0.30$, $a_2 = -0.30$, $a_1 = 0.52$, $a_3 = -0.52$
- $b_2 = 0.60$, $b_3 = 1.04$

Figure 184

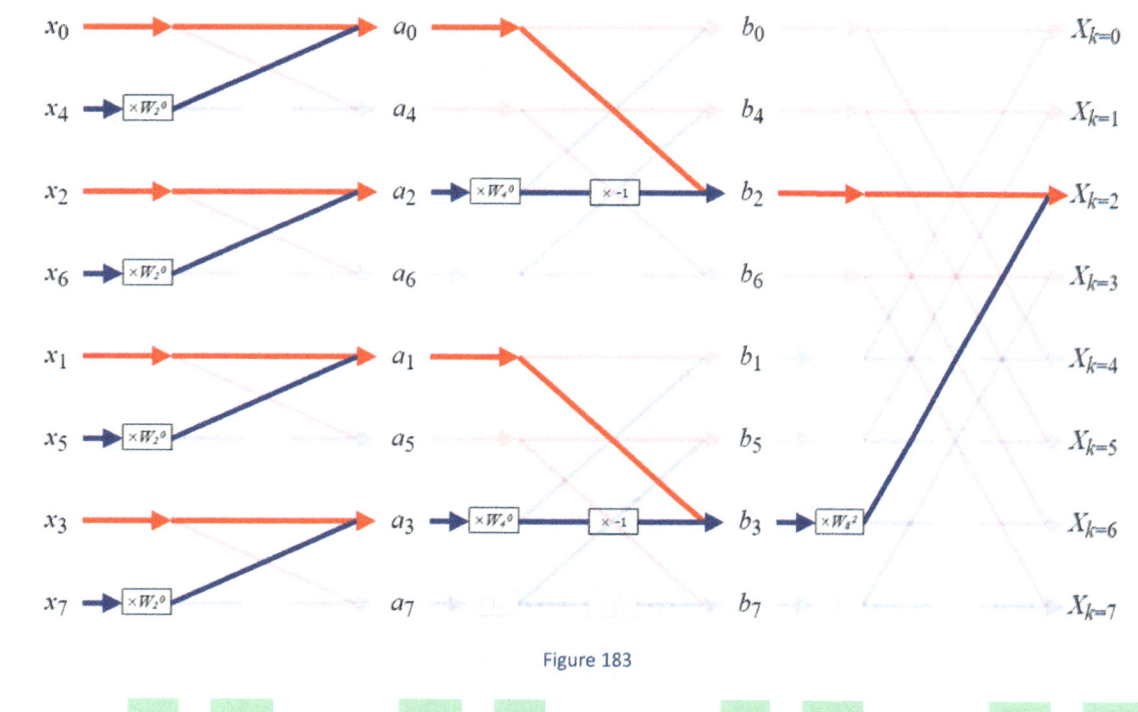

Twiddle factor W_4^0

In the 4-point DFT stage, the second sample in each pair is multiplied by W_4^0. We found out when we were calculating the DFT for frequency index $k=0$ that this twiddle factor is equal to 1. Therefore:

$$b_2 = a_0 - W_4^0 \times a_2 = 0.30 - 1 \times -0.30 = 0.60$$

Equation 105

$$b_3 = a_1 - W_4^0 \times a_3 = 0.52 - 1 \times -0.52 = 1.04$$

Equation 106

Twiddle factor W_8^2

$$W_8^2 = \cos\left(2\pi \cdot \tfrac{2}{8}\right) - i \cdot \sin\left(2\pi \cdot \tfrac{2}{8}\right)$$
$$= \cos\left(\tfrac{\pi}{2}\right) - i \cdot \sin\left(\tfrac{\pi}{2}\right)$$
$$= -i$$

Equation 107

Completing the calculation

$$X_{k=2} = b_2 + W_8^2 \times b_3$$
$$= 0.60 + -i \times 1.04$$
$$\boxed{X_{k=2} = 0.60 - 1.04i}$$

Equation 108

Calculating the DFT of frequency index *k*=6

Once again, having calculated the DFT for frequency index *k*=2, we get the DFT of frequency index *k*=6 almost for free.

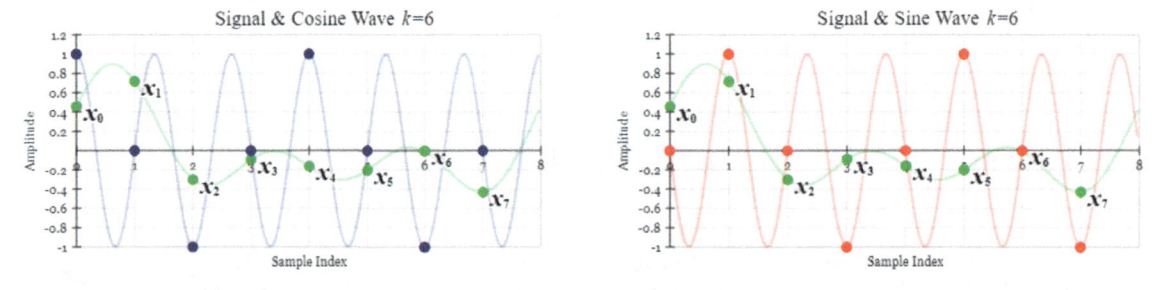

Figure 185

The DFT for frequency index *k*=6 can be computed using almost the same butterfly diagram as we used to calculate the DFT for frequency index *k*=2. The only difference is we use the result from the lower branch of the final butterfly in the 8-point DFT stage, instead of the upper branch.

How the Fourier Transform works   Chapter 5

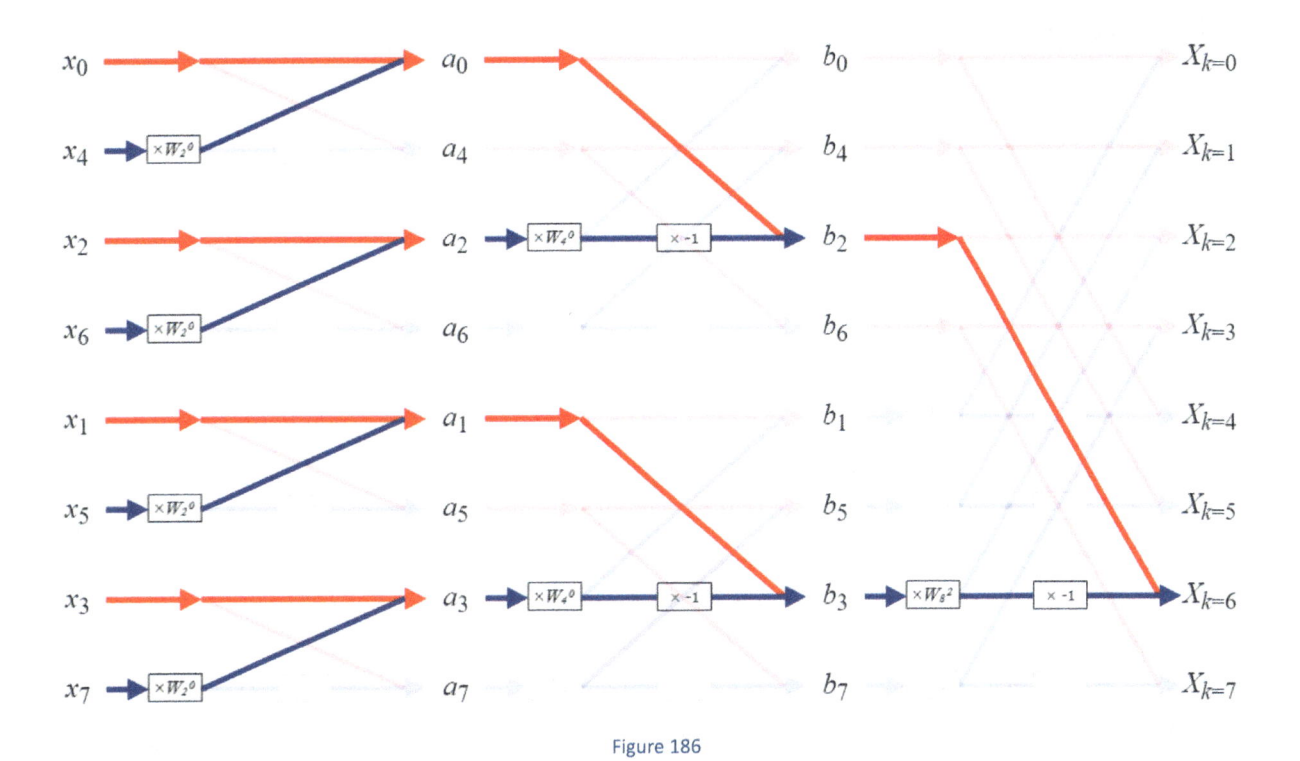

Figure 186

We can thus easily calculate the DFT for this frequency by making only one minor change to the equation. Instead of adding b_3 to b_2, we subtract it.

$$X_{k=6} = \left[(x_0 + x_4) - W_4^0 \cdot (x_2 + x_6) \right] - W_8^2 \cdot \left[(x_1 + x_5) - W_4^0 \cdot (x_3 + x_7) \right]$$

$$= (a_0 - W_4^0 \cdot a_2) - W_8^2 \cdot (a_1 - W_4^0 \cdot a_3)$$

$$= b_2 - W_8^2 \cdot b_3$$

Figure 187

Completing the calculation

$$X_{k=6} = b_2 - W_8^2 \times b_3$$
$$= 0.60 - (-i) \times 1.04$$
$$\boxed{X_{k=6} = 0.60 + 1.04i}$$

Equation 109

Calculating the DFT of frequency index k=3

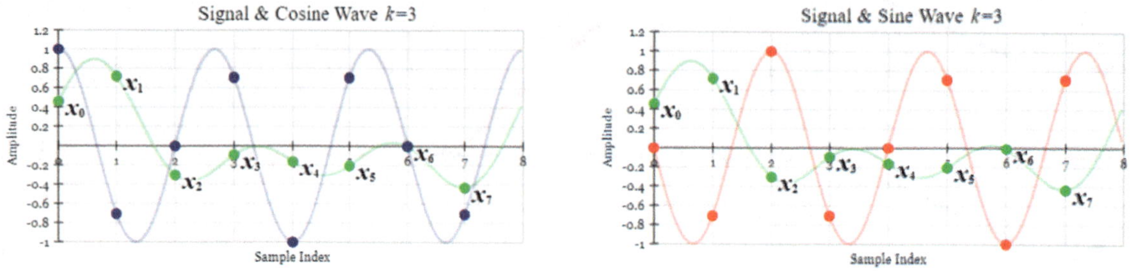

Figure 188

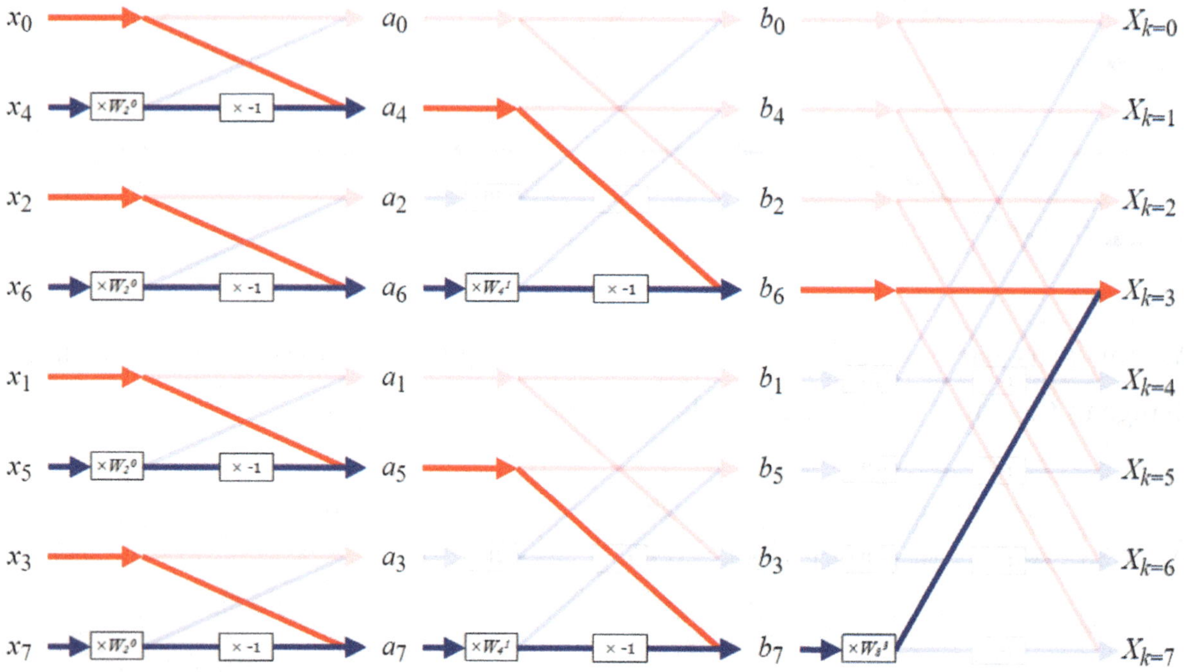

Figure 189

$$X_{k=3} = \Big[\big(\underset{0.46}{x_0} - \underset{-0.16}{x_4}\big) - W_4^1 \cdot \big(\underset{-0.30}{x_2} - \underset{0.00}{x_6}\big)\Big] + W_8^3 \cdot \Big[\big(\underset{0.72}{x_1} - \underset{-0.20}{x_5}\big) - W_4^1 \cdot \big(\underset{-0.09}{x_3} - \underset{-0.43}{x_7}\big)\Big]$$

$$= \Big(\underset{0.62}{a_4} - W_4^1 \cdot \underset{-0.30}{a_6}\Big) + W_8^3 \cdot \Big(\underset{0.92}{a_5} - W_4^1 \cdot \underset{0.34}{a_7}\Big)$$

$$= \underset{0.62\,-\,0.3i}{b_6} + W_8^3 \cdot \underset{0.92\,+\,0.34i}{b_7}$$

Equation 110

Twiddle factor W_4^1

We calculated this twiddle factor before, when the frequency index k was 1. We found out that $W_4^1 = -i$. Therefore, we can calculate b_6 as shown in Equation 111, and b_7 as shown in Equation 112.

$$\begin{aligned} b_6 &= a_2 - W_4^1 \times a_6 \\ &= 0.62 - (-i) \times -0.30 \\ &= 0.62 - 0.30i \end{aligned}$$

Equation 111

$$\begin{aligned} b_7 &= a_5 - W_4^1 \times a_7 \\ &= 0.92 - (-i) \times 0.34 \\ &= 0.92 + 0.34i \end{aligned}$$

Equation 112

Twiddle Factor W_8^3

$$\begin{aligned} W_8^3 &= \cos\left(2\pi \cdot \tfrac{3}{8}\right) - i \cdot \sin\left(2\pi \cdot \tfrac{3}{8}\right) \\ &= \cos\left(\tfrac{3\pi}{4}\right) - i \cdot \sin\left(\tfrac{3\pi}{4}\right) \\ &= -0.707 - 0.707i \end{aligned}$$

Equation 113

Complex multiplication

$$W_8^3 \times b_3 = (-0.707 - 0.707i) \times (0.92 + 0.34i)$$

First: $-0.707 \times 0.92 = -0.65$
Outside: $-0.707 \times 0.34i = -0.24\,i$
Inside: $-0.707i \times 0.92 = -0.65\,i$
Last: $-0.707i \times 0.34i = -0.24\,i^2$
$ = -0.24 \times -1$
$ = 0.24$

$$\begin{aligned} W_8^3 \times b_3 &= -0.65 - 0.24i - 0.65i + 0.24 \\ &= -0.65 + 0.24 - 0.24i - 0.65i \\ &= -0.41 - 0.89i \end{aligned}$$

Equation 114

Completing the calculation with complex addition

$$\begin{aligned} X_{k=3} &= b_6 + W_8^3 \times b_7 \\ &= (0.62 - 0.3i) + (-0.41 - 0.89i) \\ &= 0.62 + 0.41 - 0.3i - 0.89i \\ &= \boxed{X_{k=3} = 0.21 - 1.19i} \end{aligned}$$

Equation 115

Calculating the DFT of frequency index *k*=7

Once again, having calculated the DFT for frequency index *k*=3, we get the DFT of frequency index *k*=7 almost for free.

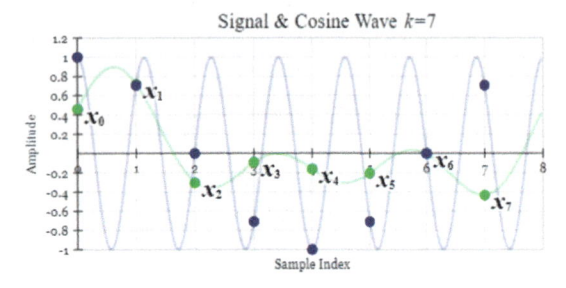

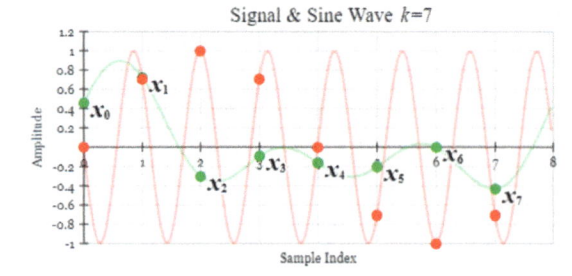

Figure 190

The DFT for frequency index *k*=7 can be computed using almost the same butterfly diagram as we used to calculate the DFT for frequency index *k*=3. The only difference is we use the result from the lower branch of the final butterfly in the 8-point DFT stage, instead of the upper branch.

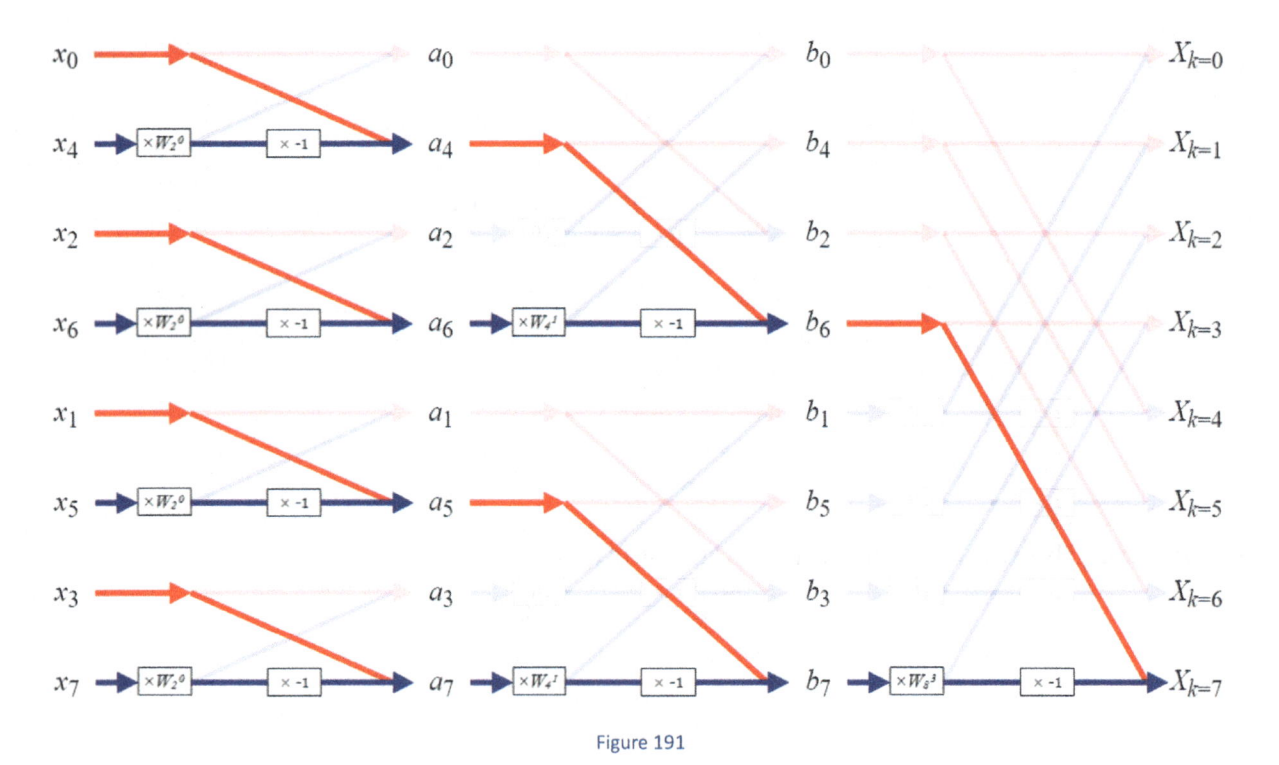

Figure 191

$$X_{k=7} = \left[(x_0 - x_4) - W_4^1 \cdot (x_2 - x_6) \right] - W_8^3 \cdot \left[(x_1 - x_5) - W_4^1 \cdot (x_3 - x_7) \right]$$

(with values: $x_0=0.46$, $x_4=-0.16$, $x_2=-0.30$, $x_6=0.00$, $x_1=0.72$, $x_5=-0.20$, $x_3=-0.09$, $x_7=-0.43$)

$$= (a_4 - W_4^1 \cdot a_6) - W_8^3 \cdot (a_5 - W_4^1 \cdot a_7)$$

(with values: $a_4=0.62$, $a_6=-0.30$, $a_5=0.92$, $a_7=0.34$)

$$= b_6 - W_8^3 \cdot b_7$$

(with values: $b_6 = 0.62 - 0.3i$, $b_7 = 0.92 + 0.34i$)

Equation 116

Completing the calculation

We can thus easily calculate the DFT for this frequency by making only one minor change to the equation. Instead of adding b_7 to b_6, we subtract it.

$$\begin{aligned} X_{k=7} &= b_6 - W_8^3 \times b_7 \\ &= (0.62 - 0.3i) - (-0.41 - 0.89i) \\ &= 0.62 + 0.41 - 0.3i + 0.89i \\ &= \boxed{X_{k=7} = 1.03 + 0.59i} \end{aligned}$$

Equation 117

Making sense of the results

The FFT tells us the properties of each sinusoid making up our signal. Sinusoids have three properties: frequency, magnitude, and phase. Therefore, for the signal we have been working with, we would expect the output of the FFT to be a table containing 8 rows (1 row per sinusoid), and three columns (frequency, magnitude, and phase).

However, as you may have noticed, the output is just the list of complex numbers shown in Table 5.

To get the information we desire, a little more work is required.

Frequency Index	FFT Output
k=0	0
k=1	1.03 – 0.59i
k=2	0.6 - 1.04i
k=3	0.21 - 1.19i
k=4	0
k=5	0.21 + 1.19i
k=6	0.6 - 1.04i
k=7	1.03 – 0.59i

Table 5

https://howthefouriertransformworks.com/

Frequency

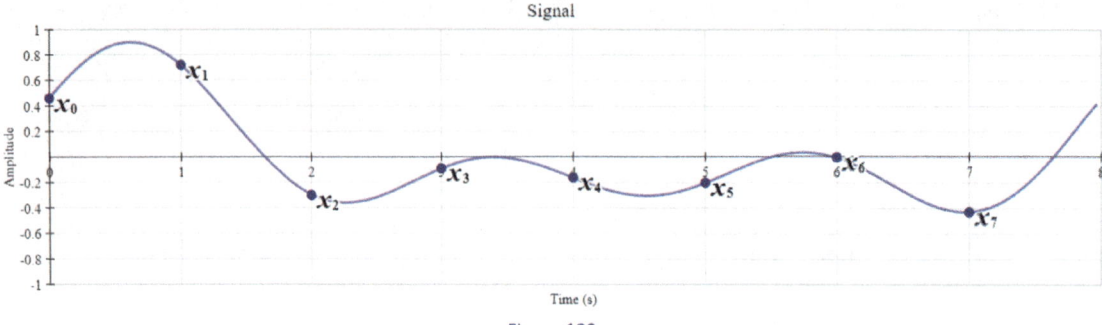

Figure 192

To calculate the frequency for each frequency index, we need to know the sampling rate. The sampling rate isn't obvious from Figure 192, as we have only plotted the sample index on the x-axis rather than the actual time in seconds. However, let's assume that we sampled this signal over 1 second. This means that the sampling rate, *R*, is 8 Hz.

We already know how many samples there are in our FFT, so *N*=8. Therefore, we can calculate the frequency (in Hz) for each frequency index *k*, using Equation 118.

$$f = R \times \frac{k}{N}$$

Equation 118

Magnitude

As we saw in Chapter 2, the magnitude, *M*, of the sinusoid for each frequency can be calculated by using the Pythagorean theorem on the real (cosine), $\Re$, and imaginary (sine), $\Im$, components of each of the FFT's output values, as shown in Equation 119.

$$M = \sqrt{\Re^2 + \Im^2}$$

Equation 119

Phase

The Fourier Transform provides us with Phase information about each sinusoid making up our signal which we can calculate by using the inverse tangent function as we discussed in Chapters 2 and 3. However, in the FFT, we have a practical form of the Fourier Transform that can be applied to many different types of signals. The problem with the inverse tangent function is that it only tells us the phase from -90° to +90°. If the phase lies outside this range, the inverse tangent function will reflect the phase value in the y-axis so that it lies within the -90° to +90° range.

For example, Figure 193 shows a sinusoid with a phase of 45°. Figure 194 shows the same sinusoid represented in the complex plane with the amplitude of its real (cosine) component plotted on the x-axis and the amplitude of its imaginary (sine) component plotted on the y-axis. The phase of this sinusoid, 45°, lies in Quadrant 1 of the graph in Figure 194.

Figure 193

Figure 194

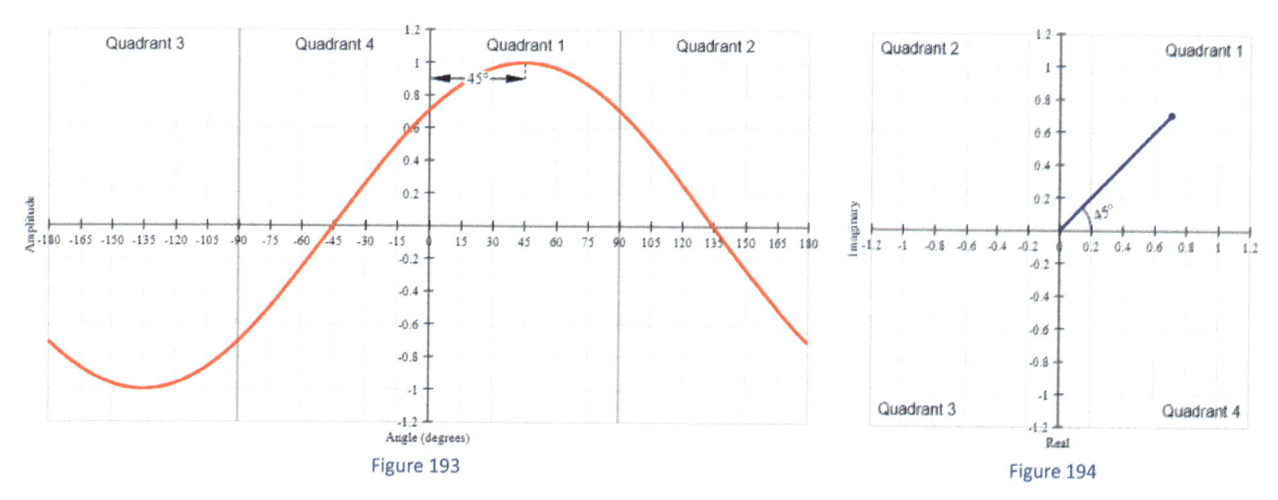

Figure 195 shows a different sinusoid with a phase of 135°. Figure 196 represents this sinusoid in the complex plane. The phase of this sinusoid, 135°, lies in Quadrant 2 of the graph in Figure 196.

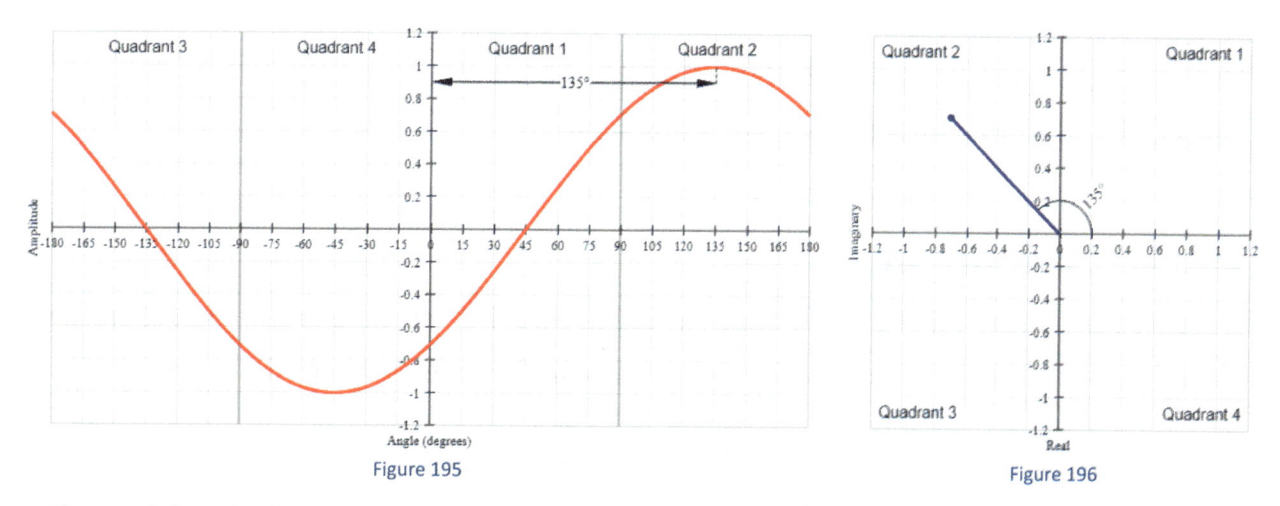

Figure 195

Figure 196

The problem is that, to the inverse tangent function, both these sinusoids look like the sinusoid in Figure 193. If we were to calculate the inverse tangent of the real and imaginary coordinates of the point in Figure 196, it would give us a phase angle of 45°.

The actual phase angle of 135°, which should be in Quadrant 2, has been reflected in the y-axis so that it appears to the inverse tangent function as 45°, which is in Quadrant 1.

In many practical situations, where we have no prior information about the phase content of a signal, we cannot assume that the phase of individual sinusoids within the signal lies within the -90° to 90° range.

In stereo audio processing for example, the phase relationship between left and right channels is crucial for creating spatial effects like panning, stereo widening, and binaural audio. Accurate, full range phase measurement is critical to maintain the intended spatial positioning of sound sources.

Therefore, to calculate the phase of the sinusoid for each frequency, we use a 4-quadrant inverse tangent function, often known as the *atan2* function. This differs from a regular inverse tangent function by remembering the signs of the real and imaginary values, as shown in Equation 120.

$$atan2\left(\frac{\Im}{\Re}\right) = \begin{cases} tan^{-1}\left(\frac{\Im}{\Re}\right) & \text{if } \Re > 0 \\ tan^{-1}\left(\frac{\Im}{\Re}\right) + \pi & \text{if } \Re < 0 \text{ and } \Im \geq 0 \\ tan^{-1}\left(\frac{\Im}{\Re}\right) - \pi & \text{if } \Re < 0 \text{ and } \Im < 0 \\ +\frac{\pi}{2} & \text{if } \Re = 0 \text{ and } \Im > 0 \\ -\frac{\pi}{2} & \text{if } \Re = 0 \text{ and } \Im < 0 \\ \text{undefined} & \text{if } \Re = 0 \text{ and } \Im = 0 \end{cases}$$

Equation 120

The inverse tangent function calculates the phase in radians. This can be converted into degrees using Equation 121.

$$\theta^o = 180 \times \frac{\theta^c}{\pi}$$

Equation 121

The FFT for our signal

We have finally arrived at our destination. Using the FFT, we have found the properties of all the sinusoids that make up our signal. They are listed in Table 6.

Frequency Index	FFT Output	Frequency	Magnitude	Phase
k=0	0	0 Hz	0.00	0°
k=1	1.03 – 0.59i	1 Hz	1.19	-30°
k=2	0.6 - 1.04i	2 Hz	1.20	-60°
k=3	0.21 - 1.19i	3 Hz	1.21	-80°
k=4	0	4 Hz	0.00	0°
k=5	0.21 + 1.19i	5 Hz	1.21	80°
k=6	0.6 - 1.04i	6 Hz	1.20	60°
k=7	1.03 – 0.59i	7 Hz	1.19	30°

Table 6

Figure 197 – Magnitude Graph

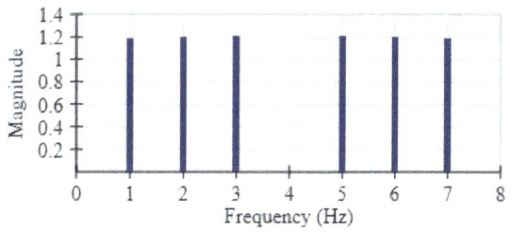

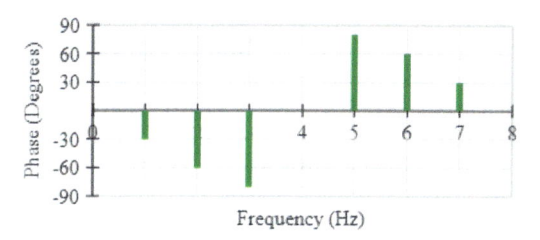

Figure 198 – Phase Graph

You might notice a certain symmetry in the magnitude and phase results. To find out why this happens scan the QR code or click on the link:

Why is the output of the FFT symmetrical?

Chapter 6: Fourier's Legacy

By looking at the individual sinusoids that make up a signal the Fourier Transform allows us to get more out of the data we collect. Using the Fourier Transform, we can design filters, build compression algorithms, keep machines running safely, recognize faces, locate the position of objects, tune musical instruments, communicate over great distances, diagnose medical problems, talk to our computers, locate faults in underground cables, design circuit boards, and much more. The list is endless. However, Fourier wasn't the only person to realize the potential of looking at signals in the frequency domain.

The Laplace Transform, and its discrete-time counterpart, the Z-Transform, are used extensively in digital filter design. Unlike Fourier, who built his signals out of sinusoids with constant amplitude, Laplace built his signals out of decaying sinusoids. This enabled the Laplace transform to model certain signals that Fourier could not.

The Wavelet Transform is used when we want to analyze data containing frequency elements that come and go from a signal. A wavelet is a brief burst of sine wave. Whereas the Fourier Transform looks at which frequencies exist over the course of an entire signal, the Wavelet Transform will look at a signal for only a short period. This enables it to locate when certain frequency patterns begin and end in a signal, something the Fourier Transform can do only if we chop up a signal into bits, with all the problems that involves, as we discussed in Chapter 4.

All these methods have one thing in common: they all look at signals, not as random events, but as a collection of sinusoids. Perhaps Fourier wasn't so far from the truth when he claimed that:

> **Any function of a variable, whether continuous or discontinuous, can be expanded as a series of sines of multiples of the variable.**

After all, it is thanks to this 200-year-old idea that we can now understand our data better than ever before.

Made in the USA
Las Vegas, NV
22 September 2023